MICROBIOLOGY AND MOLECULAR DIAGNOSIS IN PATHOLOGY

MICROBIOLOGY AND MOLECULAR DIAGNOSIS IN PATHOLOGY

A Comprehensive Review for Board Preparation, Certification and Clinical Practice

AUDREY WANGER

University of Texas McGovern School of Medicine, Houston, TX, United States

VIOLETA CHAVEZ

University of Texas McGovern School of Medicine, Houston, TX, United States

RICHARD S.P. HUANG

Roche Diagnostics, Tucson, AZ, United States

AMER WAHED

University of Texas McGovern School of Medicine, Houston, TX, United States

JEFFREY K. ACTOR

University of Texas McGovern School of Medicine, Houston, TX, United States

AMITAVA DASGUPTA

University of Texas McGovern School of Medicine, Houston, TX, United States

Elsevier
Radarweg 29, PO Box 211, 1000 AE Amsterdam, Netherlands
The Boulevard, Langford Lane, Kidlington, Oxford OX5 1GB, United Kingdom
50 Hampshire Street, 5th Floor, Cambridge, MA 02139, United States

British Library Cataloguing-in-Publication Data
A catalogue record for this book is available from the British Library

Library of Congress Cataloging-in-Publication Data
A catalog record for this book is available from the Library of Congress

ISBN: 978-0-12-805351-5

For Information on all Elsevier publications
visit our website at https://www.elsevier.com/books-and-journals

Publisher: Mica Haley
Acquisitions Editor: Tari Broderick
Editorial: Kathy Padilla
Production: Karen East and Kirsty Halterman
Designer: Greg Harris

Typeset by MPS Limited, Chennai, India

CONTENTS

PREFACE

Preparing for examination conducted by American Board of Pathology is an arduous task as residents must have mastery in all subspecialities of clinical pathology including clinical chemistry, point of care testing, hematology, coagulation, microbiology, blood banking as well as laboratory management including quality control and statistics. The first book to help pathology residents to prepare for the board; "Clinical Chemistry, Immunology and Laboratory Quality Control: A Comprehensive Review for Board Preparation, Certification and Clinical Practice" was published, by Elsevier in January 2014. After receiving favorable reviews we wrote the second book in the series: "Hematology and Coagulation: A Comprehensive Review for Board Preparation, Certification and Clinical Practice" which was also published by the Elsevier in February 2015. That book was also well received by pathology residents and others.

Success of the first two books in this series encourages us to write the third book "Microbiology and Molecular Diagnostics in Pathology: A Comprehensive Review for Board Preparation, Certification and Clinical Practice" in a series of books, designed for board review for Pathology residents. This book is not an alternative to any standard microbiology textbook but a compendium to help the resident prepare for the pathology board examination. This book has 13 chapters and provides a comprehensive review of all important aspects of microbiology including molecular diagnostics and a chapter on management of the microbiology laboratory. We deliberately avoid figures in this book similar to our previous books because such figures are available in standard text books. Our goal is to allow the final cost of the book to remain reasonable. We have added a section, denoted as "key points" at the end of each chapter. We hope that this section will be a good resource for reviewing information, when time at hand is somewhat limited just before taking the board examination. Moreover, we are confident that medical students with a keen interest in microbiology as well as fellows in microbiology will find this book helpful. In addition, family medicine and internal medicine doctors may also find this book useful as a source of comprehensive review of microbiology and molecular diagnostics from pathologist's standpoint. We would like to thank our Department Chair, Dr. Robert Hunter for encouraging us to write the third book in the series and his support during our long process of preparing the manuscript. If pathology residents and other readers find this book useful, our efforts will be dully rewarded.

Respectfully submitted by
Audrey Wanger, Violeta Chavez, Richard Huang, Amer Wahed, Jeffrey K. Actor and Amitava Dasgupta

CHAPTER 1

Laboratory Management and Administration

Contents

INTRODUCTION

Clinical laboratories which are involved in analyzing specimens from patients for diagnostic purposes are required by law to have proper standard of practice and accreditation from a suitable agency and for some states from the state regulatory agency. This chapter will discuss issues with quality control (QC), biosafety, College of American Pathologists (CAP), and Clinical Laboratory Improvement Amendments (CLIA) regulation essential to maintain a certified clinical microbiology laboratory in operation [1].

CLINICAL LABORATORY IMPROVEMENT AMENDMENTS/ COLLEGE OF AMERICAN PATHOLOGISTS AND CLINICAL LABORATORIES

Clinical laboratories are regulated by Center for Medicare and Medicaid Services (CMS) through the CLIA, whose objective is to assure a standard of quality in laboratory testing. The CLIA'88 Law, in effect since 1992, requires all clinical laboratories doing testing on human specimens to be certified by CMS. The requirements vary depending on the complexity of testing performed by the laboratory. Complexity varies from waived testing which includes rapid strep and other antigen testing to moderate complex tests for example the Gram stain and urine culture to high

Microbiology and Molecular Diagnosis in Pathology.
DOI: http://dx.doi.org/10.1016/B978-0-12-805351-5.00001-6

complexity tests which includes most other testing in the microbiology laboratory. Laboratories that only perform waived testing are eligible for a certificate of waiver under CLIA. These laboratories receive no routine inspections although a small percentage will be subjected to a random inspection to assure quality.

All laboratories that perform moderate or high complexity testing must meet certain standards for proficiency testing (PT), QC, quality assurance (QA), and personnel qualifications. In order to assess laboratories compliance with these regulations laboratories must be inspected every 2 years by either CMS or an organization approved by CMS most commonly CAP or The Joint Commission (TJC). CAP inspection process surveys all areas of the laboratory and is a peer review process. TJC accredits hospitals and usually performs an unannounced inspection every 3 years. Although a hospital laboratory is inspected by the CAP, the hospital inspection by TJC can involve the laboratory. Their inspection typically use a tracer process which follows a laboratory test through the process of collection, transport, testing, and observing the integration of laboratory processes throughout the hospital.

Qualifications are specified for personnel, including the Lab director, technical consultant, clinical consultant, and testing personnel. These qualifications are based on the level of testing complexity performed. There are no personnel requirements for waived testing; however, there are very specific requirements for labs performing moderate or high complexity testing. The most common organization providing certification for testing personnel is the American Society of Clinical Pathology (ASCP). Requirements can vary depending on the state in which the personnel practice and their discipline such as transfusion medicine which has additional requirements.

QUALITY ASSURANCE AND MANAGEMENT

All laboratories are required by CLIA to have a QA program. This program is intended to establish a plan to identify and correct problems, assure accuracy, and timeliness of results reported and to assure competency of testing personnel. Quality goals are set based on the needs of the individual facility. A system must be in place to determine if the goals are met and corrective action to be taken if they are not met.

The Clinical Laboratory Standards Institute (CLSI) is a nonprofit voluntary organization that is charged with development of consensus guidelines for clinical laboratories. Examples of consensus documents for microbiology include guidelines for identification and susceptibility testing of organisms. This document recommends antimicrobials to be reported against particular organisms causing different infections and provides breakpoints for susceptibility of antimicrobials against a broad range of routine and fastidious bacteria and fungi. Documents are also available on topics such as writing a procedure manual, QA, biosafety in the laboratory and verification and validation of processes. These documents are updated yearly, every 3 years, or as needed. These documents are a guide to help laboratories in establishing their quality program.

Patient test management is an important part of the QA program. Included is the assessment of all processes, preanalytical which includes all processes prior to testing, analytical, which is the testing portion, and postanalytical, which includes review of results. Procedures to assure appropriate specimen collection and transport to the laboratory, relevance and necessity of testing, specimen rejection and cancellation if testing is deemed inappropriate, and timely and accurate reporting of results must be in place. All laboratories must have a manual detailing proper specimen collection and transport that is available to the users of the laboratory. This document must include criteria for specimen rejection and policies on the process of specimen rejection. Procedures for specimen retention, storage, and retrieval must also be appropriate for the laboratory and the testing performed. Laboratories that send testing to a reference laboratory should assure the quality of those test results as well by assuring that the reference laboratory also participates in a QA program.

Competency of personnel is required and is part of the assessment process for CAP and TJC as well as others. Competency assessment of all procedures and processes including preanalytical, analytical, and postanalytical phases of these processes is mandatory for all testing personnel. It is the responsibility of the laboratory director to ensure that procedures are in place that assesses all these phases including specimen processing, test performance, and result reporting. This must occur at least semiannually during the initial year of employment, and at least annually thereafter or when significant changes to a process have been implemented. Competency includes direct observation of processes, monitoring of results reported, unknown testing (PT), and testing problem solving ability. Processes must also be in place to re-train employees whose competency was not successful.

A manual containing procedures for all technical processes must be available to the staff performing testing. The presence of a package insert is not sufficient for a laboratory procedure. Evidence must be demonstrated that all staff have read the procedures and are aware of any modifications made to the procedures. The laboratory director is responsible for assuring that all procedures are technically appropriate and up to date for the testing performed by reviewing every 2 years or when significant changes are made. A current CAP checklist can provide guidance for the contents of the procedure manual including standards for workup of different types of cultures.

PT is typically accomplished by enrolling in a program provided by the CAP or a comparable alternate program. A laboratory must perform PT for each test performed on patients. If no PT is available from CAP alternate testing can be performed by sharing samples with another laboratory. Testing of proficiency samples must be performed using the same methods and worked up to the same extent as patient samples with the exception of not sending any PT samples out to a reference lab. CAP may mandate the laboratory cease testing after successive PT failures which can be reversed after acceptable corrective action.

A problem log documenting issues noted in the laboratory or those identified through physician complaints must be maintained. A process must also be in place to address these issues. Root cause analysis (RCA) is a comprehensive, systematic analysis for identifying causal factors underlying variation in performance including occurrence of a sentinel event. A sentinel event as defined by TJC is a patient safety event that results in serious harm or death. The RCA is designed to understand the cause of variation in processes that may have led to a serious error without placing blame on an individual and to suggest ways to prevent recurrence of that event.

QC is necessary to ensure that all tests are working the way they are intended every time the test is performed. Procedures must be in place noting the type and frequency necessary for testing as well as corrective action needed to remedy incorrect results. QC must be performed on media used in the microbiology laboratory. The media manufacturer is required to provide the lab with documentation of QC. Most media are exempt from further testing, but a list of media that require further testing by the laboratory is provided in the CLSI document. Even though the manufacturer provides documentation of QC having been performed the laboratory is still responsible for documentation that the media has not been compromised in storage and or shipment to the laboratory from the distributer. This includes checking all media for damage or contamination. In addition QC is required to be performed on all reagents, stains, rapid identification kits, and automated instruments used by the lab in an interval deemed appropriate by either the manufacturer or CLIA. This is typically once per day or week but can be as frequent as every time a test is run. If multiple test methods are used to perform the same process either manual or automated a comparison must be made on an on-going basis to assure that results are equivalent.

Recently, CLIA implemented a new program called individualized quality control plan (IQCP) in an attempt to reduce unnecessary QC and to individualize the frequency of QC required for each test or process based on the likelihood of a QC failure. Reducing QC is appropriate as long as a failure does not increase the likelihood of causing harm to a patient. As of 2016, laboratories had to choose implementation of IQCP for a test process or continue to perform QC at the frequency required by CLIA. The IQCP can be used on most nonwaived testing and includes a risk assessment, a QC plan to document processes performed to decrease risk of test failure, and a QA plan to monitor on an on-going basis the effectiveness of the QC plan. The IQCP is an individualized program tailored to the needs of the individual lab. A risk assessment must be performed for each test based on the laboratories own data. Computerized programs are available to aide laboratories in creating an appropriate plan.

Another significant part of the QA process is the validation and verification of new tests brought online by the laboratory and on-going review of all tests to assure accuracy in patient testing. Verification is a one-time process performed prior to patient testing to ensure that a test is performing in the testing laboratory as expected according to the manufacturer's reports (as stated on the package insert). This includes sensitivity and specificity data. Validation is an on-going process to insure that the test continues to perform as expected. Validation is performed by using the routine laboratory processes in place including QC, PT, and competency testing, but must be documented.

Some important terms are as follows:

- Sensitivity is a measure of how well a test can accurately detect patients with a particular disease.
- Specificity is a measure of how accurately a test can identify all noninfected patients.
- Positive predictive value is the probability that a positive result accurately indicates the presence of a specific disease.
- Negative predictive value is the probability that a negative result accurately indicates absence of a specific disease.
- The predictive value is dependent on the prevalence of the disease in the population.

Screening tests should have high, greater than 95% sensitivity, and negative predictive value so that a negative test assures with great confidence that the patient does not have the disease. Confirmation tests may have decreased sensitivity, but higher specificity than screening tests.

Verification is required on all Food and Drug Administration (FDA) cleared assays of either moderate or high complexity test, even when the laboratory replaces a test system or instrument (with the same model or a different model), adds a new test, or changes the manufacturer of a test system. New tests must be checked for accuracy and precision (reproducibility). In order to determine accuracy of a new test it must be compared to a reference method or gold standard or to the comparable test the lab is currently using and looking to replace. Precision is assessed by testing samples in duplicate or triplicate and if a manual method of testing is performed its precision should be determined by different technologists to exclude the presence of operator error. For quantitative tests the reportable range must also be compared to manufacturer's specifications by testing samples that fall in different areas of the range. The criterion for acceptable comparative data varies and depends on the likelihood that discrepant results could affect care of the patient. For example, comparing different antimicrobial susceptibility systems in the lab necessitates calculating major and minor errors. Assuring that the very major errors (e.g., report of susceptible when an isolate is truly resistant to a particular antimicrobial) are minimal, and that nearly 100% categorical agreement is achieved (e.g., all susceptible isolates are susceptible and resistant isolates resistant).

As mentioned above, validation is an on-going process that by means of QC testing, PT, and competency checks of all testing personnel assures the continued accuracy of testing. In addition, laboratories that use multiple instruments to perform similar testing must compare results from the different instruments at least twice per year. For example, if multiple instruments (either the same manufacturer or different) are used to perform antimicrobial identification and susceptibility the lab must assure that all instruments provide comparable results. This can be achieved by rotating the daily or weekly QC testing on all instruments as well as challenging the instruments with stock organisms. If laboratories under the same number perform the same testing at different locations, comparisons must also be made between the results from all instruments.

The QA program must include a process for monitoring quality indicators. If an IQCP was implemented, a plan must be in place for ongoing review of its effectiveness. Blood culture contamination rates are a common quality indicator that is monitored in the microbiology laboratory. Laboratories can benchmark their data against national standards or specifically against hospital laboratories most similar to themselves.

COMPLIANCE

The Health Insurance Portability and Accountability Act (HIPAA) of 1996 regulates the way in which health information is handled and processed providing patient confidentiality and protection of their health care information. Patients must have access to their health information and be protected from inappropriate use of that information. Therefore a patient must consent to use of their protected health information (PHI). A business associate agreement must be in place in order for a third party to handle PHI. A third party can be a consultant helping with accounting or other processes medical or financial.

A hospital information system (HIS) and a laboratory information system (LIS) are critical components of the working of the clinical laboratory. These systems are necessary for entering of patient data and making it accessible to users including medical practitioners and patients. The LIS can also be useful for automated retrieval of QC information and maintenance of these critical records. Patient test reports, instrument printouts, maintenance of records, QC and QM records, PT, and competency records must all be maintained for at least 2 years. An important component of both the HIS and LIS is that patient confidentiality is maintained during data retrieval.

The LIS is also helpful for document control which is required to assure that all laboratory procedures and policies are maintained in an easily accessible location, typically electronically. The LIS may also help in specimen tracking for easier results retrieval and also in specimen retrieval for further additional testing or review when required. Patient specimens should be retained for at least 48 hours with the exception of urines which only needs to be kept for 24 hours. Gram stain slides prepared from positive blood culture bottles needs to be kept for at least 7 days.

LOINC: Logical observation identifiers names and codes is a database of universal codes to identify laboratory observations that allow for measurement of quality indicators and outcomes management. Health care facilities are required to use LOINC codes to transfer reportable disease information to public health laboratories. CMS has incentivized laboratories to encourage use of electronic medical records which has the ability to share information across facilities.

BUDGET

There are different types of laboratory budgets most commonly referred to as an operating budget. The operating budget includes forecast activity for the unit and gross charges that will be generated by the forecasted volume. It also includes an expense budget which projects amount of resources that will be required to produce the forecasted volumes. Particularly important are the number of full time equivalents (FTEs) required to perform the testing. An FTE can be made up of one or more employees paid for a total of 2080 hours per year based on 8 hours per day × 5 days per week × 52 weeks per year. Based on a combination of productive and nonproductive time (vacation and sick leave) a laboratory would need 1.1 or more FTEs to actually have 1.0 FTE of productive time. The other part of the operating budget is the capital budget which includes new or replacement property that will be used for at least a year and overhead and expenses such as utilities, building maintenance, and hospital administration (personnel services and purchasing services). The return on investment which determines the benefit the laboratory will realize that a capital investment needs to be calculated. If an instrument is purchased as capital equipment the amount of time to make that money back must be calculated. Depreciation must also be taken into account on purchased equipment and is usually divided over the predicted life of the instrument.

LABORATORY SAFETY

All laboratories must have procedures in place for maintaining a safe working environment. There must be documentation that all employees are trained on those procedures. This includes fire safety and chemical safety. Safety training should be done for all new employees and at least annually thereafter. Universal or standard precautions must be used at all times by all personnel handling patient specimens. All personnel must have access to appropriate personal protective equipment (PPE) and be trained on their use. The Hepatitis B vaccine should be made available to all personnel who might possibly be exposed to blood and body fluids. There must be evidence that the laboratory follows up on any possible exposure to blood or body fluid through either splash or needle stick. Employees should also have access to ergonomically suitable work stations to prevent musculoskeletal damage.

A plan must also be in place for handling of agents of bioterrorism including use of a biological safety cabinet and the extent of workup that should be performed in the local laboratory. Technologists should be educated on the proper tests that can safely be used for preliminary identification of these organisms without placing them the organism in an automated instrument. Suspicious isolates which cannot be ruled out as an agent of bioterrorism should be appropriately packaged and sent to a reference lab for confirmation or exclusion. Training programs are available from local health department laboratories or online for identification and safe shipping of these and other infectious specimens. A plan should also be in place for containment of spills in the laboratory either chemical or biological.

Laboratories must also have a tuberculosis control plan which includes testing of personnel at risk at a defined interval (6 months to 1 year depending on their risk) and training on appropriate use of biological safety cabinets and N-95 masks or respirators. Fit testing must be performed to assure appropriate use of these specific PPE.

A chemical hygiene plan must also be in place that assures all personnel are knowledgeable of the chemicals in use in their area and have access to current safety data sheets previously called MSDS. A safe environment includes as mentioned above access to biosafety cabinets that are in proper working order and are inspected yearly. Fire extinguishers, blankets, eyewash stations, and spill kits are readily available to all staff. An easily accessible fire escape plan should be posted in all areas of the lab.

A plan also must be in place for safe disposal of potentially hazardous waste. Optimally all possibly infectious material generated in the microbiology laboratory, particularly the mycology and mycobacteriology laboratories should be decontaminated prior to removal from the testing area. If this is not possible procedures for safe handling of this waste prior to disposal must be in place. The CAP checklist has guidance on all of these many requirements to make it as easy as possible to maintain compliance with the regulations.

KEY POINTS

- The CLIA'88 Law, in effect since 1992, requires CMS certification of all laboratories testing human clinical specimens. Laboratories that perform moderate or high complexity testing must meet certain standards for PT, QC, and QA.

- Personnel qualifications based on level of testing complexity are required, and can depend on the state where the laboratory is located. The ASCP is the most common organization to provide certification.
- Laboratories are required by CLIA to have a QA program with goals based on individual facility needs. The CLSI is a nonprofit voluntary organization charged with development of consensus guidelines for clinical laboratory practice.
- Laboratories must have detailed manual outlining processes for QA, QC, and specimen collection. Criteria for specimen rejection must be included. Manuals must also be in place for all procedures performed, with review on a regular basis and revisions made when changes to process are implemented. Problem logs must also be maintained.
- Validation and verification of new tests, and on-going review of all current tests, must be completed on a regular basis to assure accuracy in sample testing. Screening and confirmation of sensitivity must also be addressed. New tests implemented must be compared to currently used assays, when possible. Benchmarks against national standards should be employed.
- The HIPAA regulates how health information is handled and processed, and provides patient confidentiality of related health care information. Together with the HIS and the LIS, they ensure patient competency records.
- Laboratory operating budgets forecast activity for the unit and gross charges generated, based on predicted test specimen volume. Instrumentation, and associated depreciation, is included in the budget.
- Laboratory safety procedures include those that are critical for a safe working environment. Access to PPE is required, as well as training for fire and chemical emergencies. Plans must also be in place for bioterrorism events.

REFERENCE

[1] Garcia LS, editor. Clinical laboratory management. 2nd ed. Washington, DC: American Society for Microbiology; 2014.

CHAPTER 2

Specimen Collection and Handling in Microbiology Laboratory

Contents

INTRODUCTION

Acceptable and unacceptable specimens and appropriate specimens for diagnosis of infectious diseases are reviewed in this chapter. Differentiation of normal flora versus pathogens and reporting of organisms isolated from various body sites will also be addressed.

The importance of proper specimen collection must be stressed. Appropriate specimen collection is invaluable in order for organisms that are the cause of an infection to be rapidly isolated and identified. Avoiding growth of organisms considered normal flora at the site of collection is critical in the rapid identification of the true pathogens. Normal flora will vary depending on the site of infection and specimen collection. If possible, clinical specimens should be collected prior to initiation of antimicrobial administration to prevent inhibition of pathogen growth. Most specimens from normally sterile sites can be collected directly into a sterile container. Cultures for routine organisms, anaerobes, mycobacteria, and fungi can be

Microbiology and Molecular Diagnosis in Pathology.
DOI: http://dx.doi.org/10.1016/B978-0-12-805351-5.00002-8

performed from this type of collection methodology. Samples should be transported to the laboratory and plated within 2 hours postcollection. If this is not possible, specimens should be placed into transport media to maintain viability for extended periods of time. This is especially important for growth of fastidious organisms. Culture or other testing of viruses also requires a transport media. Additional information will be included in the following sections that discuss specific specimen types. Transportation of specimens in the appropriate containers and at the appropriate temperature is also vital to ensure the viability of the organisms.

BLOOD CULTURES

Skin microbiota represents an ecosystem of living biological organisms. Proper procedures must be followed to limit contamination of resident microflora. Blood cultures should only be taken after skin decontamination with 70% alcohol followed by chlorohexidine. Bottles should be filled with 5–10 mL blood per bottle; in the case of pediatric patients the volume is based on the weight of the child. Two to three sets (aerobic and anaerobic bottle) should be collected for optimal yield in detection of bacteremia. Each set is collected as a separate stick in a separate site. Growth from blood culture bottles indicates bacteremia and can provide diagnosis for directed therapy with specificity for the specific organism. Growth is typically detected by automated instruments that both incubate bottles and measure the production of CO_2 produced by the organisms. Bottles are incubated for 5 days; most significant organisms will grow within 24–48 hours. Contamination of blood cultures with skin organisms complicates the interpretation of positive blood cultures. Organisms that typically represent contamination species include *Staphylococcus* species other than *Staphylococcus aureus*, *Corynebacterium* species, *Bacillus* species other than *Bacillus anthracis*, *Propionibacterium acnes*, *Micrococcus* species, and Viridans *streptococci*. Collection of multiple sets of blood cultures helps to rule out contamination since true positives will likely grow from more than one set of blood cultures. Patients with intravascular infections, such as endocarditis, will likely have continuous bacteremia or high grade bacteremia. In this instance, all blood cultures collected will be positive. This is in contrast to patients with abscesses that frequently result in intermittent bacteremia.

True pathogens include *S. aureus*, *Streptococcus pneumoniae*, *Escherichia coli*, other Enterobacteriaceae, *Pseudomonas aeruginosa*, and *Candida* sp.

It is less likely to be anaerobic organisms, such as *Bacteroides* species and *Clostridium* species, and other yeast and fungal organisms.

Blood cultures can also be collected to diagnose line-related infections. The traditional practice of collection of catheter tips to diagnose line-related infections has gone out of favor in comparison to collection of "quantitative" blood cultures. One set of blood cultures should be collected from the questionably infected line and a second set from the periphery at approximately the same time. Growth of organisms in the set drawn through the line at least 2 hours before the peripheral set of bottles is significant for a line-related infection [1].

BODY FLUID COLLECTION

Body fluids should be sterile. However, body cavities or secretory organs provide localized sites where organisms can flourish usually following a bacteremia or infection at the local site. Samples should be collected during the surgical procedure and not from subsequent drains that are placed during the surgical procedure. Body fluids should be transferred directly to a sterile container prior to transport. Fluids collected by aspiration with a needle and a syringe should be inoculated into a sterile cup or red top tube with no additive. A few milliliters of joint fluid can be placed into blood culture bottles at bed side to increase the sensitivity for recovering fastidious organisms such as *Kingella*. Joint infections can be caused by *S. aureus*, while arthritis can be caused by *Streptococcus pyogenes*, *Neisseria gonorrhoeae*, and some anaerobes [2,3]. Shoulder joints in particular can be infected with *P. acnes*; specimens should be inoculated into a broth that can be held anaerobically for 10 days for the growth of that organism.

Likewise, peritoneal fluid may be contaminated by gastrointestinal microbiota when an intestine is ruptured, though patients with peritonitis likely have presence of pathogens which may include *Staphylococcus* spp., Viridans group streptococci, Gram-negatives, *Candida* spp., and fungi [4].

Keep in mind that the presence of any organisms found in areas that are normally sterile are considered significant and presence of identified bacterial isolates must be reported.

Cerebral Spinal Fluid

Cerebral spinal fluid (CSF) typically represents colorless body fluid found in the spinal cord and brain. Upon collection, fluid is placed in a sterile screw cap tube. It is critical to perform skin decontamination prior to collection.

Any organisms recovered from the CSF are to be considered potential pathogens. In children under 10 years of age, Group B streptococci, *Listeria monocytogenes*, *E. coli*, *Haemophilus influenzae*, *S. pneumoniae*, and *Neisseria meningitidis* are the most common causes of acute meningitis [5].

For the rest of the population, including older adults, *N. meningitidis* and *S. pneumoniae* are the most common causes of bacterial meningitis; *L. monocytogenes* should also be considered in older adults [6].

Keep in mind that shunt infections often result in infectious agents in the CSF, typically *Staphylococcus* spp., *Corynebacterium* spp., *P. acnes*, Gram negative bacteria, or anaerobes [7].

Although typically 1 mL of CSF is sufficient to culture for bacterial meningitis culture of CSF for mycobacterial infection such as tuberculosis requires collection of at least 5–10 mL of CSF. Molecular testing of the CSF for viral pathogens typically requires 0.5 mL/test. Review of the CSF WBC count, differential, and glucose and protein along with the patient history can help to guide the clinician in the workup of meningitis/encephalitis.

SPECIMEN COLLECTION FROM GENITAL TRACT

Genital tract specimens are submitted by collection of pus or abscess material that is placed in a sterile container. The detection of sexually transmitted pathogens is usually performed by molecular methods due to the lack of sensitivity of culture for these organisms. Specimens should be collected using the appropriate device based on the particular methodology used in the local laboratory. Collection methodologies vary depending on specifics of tissue involved, but typically involved use of a syringe and needle or appropriate swab. Urine can also be collected in a sterile container for analysis of these organisms.

Specimen Collection From Female Genital Tract

For endocervix specimens, wipe cervix clean of vaginal secretion and mucus, collect a swab with speculum and without lubricant.

For specimens from the vagina, ulcerations can be tested for yeast in either a routine or fungal culture. Wet mounts can diagnose *Trichomonas* and identify clue cells. Since *Trichomonas* needs to be mobile to make a diagnosis the swab in transport media must be sent to the lab within several hours. *Trichomonas vaginalis*, *Candida albicans*, and with *Gardnerella*, the causative agent of bacterial vaginosis, can be diagnosed using DNA technology after the appropriate swab placed in appropriate transport media is sent to the lab. *Streptococcus agalactiae* (GBS) is an important organism to identify in the cervix/rectum of pregnant women prior to delivery to prevent transmission to their baby. A cervical/rectal swab should be collected in a selective media such as Lim broth which inhibits the growth of normal flora organisms and allows for the growth of small but significant numbers of GBS. Common normal flora of the vagina include lactobacilli, diphtheroids, *S. aureus*, *S. epidermidis*, *Streptococcus* spp., *E. coli*, and *Candida* spp. [8]. Diagnosis of vaginosis should not be diagnosed by culture since it is multifactorial and should therefore be diagnosed by use of a scored Gram stain [9].

Specimen Collection From Male Genital Tract

Urethra specimens are collected after wiping tissue clean with sterile gauze, followed by using a swab to collect urethral secretion or free discharge.

Pathogens of the urethra include *Chlamydia trachomatis*, *N. gonorrhoeae*, *Mycoplasma genitalium*, and *Ureaplasma urealyticum*. Specimens are collected for molecular detection of *C. trachomatis*, *N. gonorrhoeae*, and *Mycoplasma/Ureaplasma*, and are performed as described above. Collection from a genital lesion for Herpes simplex virus should be performed using a flocked swab after unroofing the lesion. The flocked swab needs to be placed directly into viral transport media.

URINE SPECIMEN COLLECTION

The cautious collection of urine is imperative due to relative ease of contamination from normal flora. Clean-voided midstream collection is highly recommended to decrease contamination.

Urinary tract infections (UTIs) are dependent on bacterial counts and diseases presentation which will determine necessity of treatment. The method for streaking urine for colony counts depends on the size of the loop, which can be 0.01 mL where one colony equals 100 CFU/mL, or 0.001 mL where one colony equals 1000 CFU/mL. Epidemiological studies show that bacterial counts of $\geq 10^5$ CFU/mL are associated with bacterial UTI [10]. Clean catch specimens should be processed with a 0.001 mL loop and if more than or equal to 10,000 colonies are counted, then organism identification and antibiotic susceptibilities should be completed. For catheter acquired specimens, a 0.01 mL loop should be used for processing; presence of 1000 colonies or greater should be worked up for organism identification and antibiotic susceptibilities [11].

The major causes of UTIs include the Gram negative bacteria *E. coli*, *Enterococcus* spp., *Klebsiella*, *Enterobacter*, and *Proteus*. Gram positives include *Enterococcus* species and *Staphylococcus saprophyticus* in young women.

Normal floras that may be found in contaminated cultures include diphtheroids, *Staphylococcus* spp., *Lactobacillus* spp., *Neisseria* spp., and anaerobes.

STOOL SPECIMEN COLLECTION

Gastroenteritis is associated with bacteria, viruses, and parasites. Clinical correlation with diarrhea associated illnesses allows for directed and adequate testing of pathogens. Duodenal aspirates can be obtained, though significant amounts of bacteria, may be associated with delayed healing and may predict defects in mobility of the intestines [9].

Microbiology laboratories typically routinely test for *Salmonella* spp., *Shigella* spp., and *Campylobacter* spp. Stool samples collected in a sterile container is an appropriate methodology unless the sample will not reach the lab before an extended period of time has elapsed; if so, then a transport media should be used. A rectal swab containing stool material is acceptable from infants, although it is not as sensitive as a larger sample collected in a cup. Culture is useful for diagnosis of acute gastroenteritis in a patient who has not been in the hospital for more than 3 days. Patients hospitalized for more than 3 days have usually been given antibiotics and diarrhea that develops in the hospital is most likely due to *Clostridium difficile*. Stool samples for detection of *C. difficile* toxin must be liquid samples, which take the shape and size of the container. Testing is performed using molecular detection of toxin, and should not be repeated within 7 days or used as a test of cure.

Stool samples for parasites should be collected in a transport jar specific for suspected parasites. A direct smear and concentrated smear can then be performed to look for organisms. Some parasites can also be detected by methods that look for antigen of the organism which may be more sensitive than using microscopy. Fresh stool collected in a sterile cup is required for these tests, including a test for *Cryptosporidium*. Viral antigens can also be detected in fresh stool specimens. Newer technologies for the rapid diagnosis of gastroenteritis are including multiplex polymerase chain reaction (PCR), which is discussed in a subsequent chapter.

SPECIMEN COLLECTION FROM RESPIRATORY TRACT

Culture of the upper respiratory tract for the diagnosis of pharyngitis is made using a throat swab looking for *S. pyogenes* or GAS, as the primary pathogen. Throat swabs can also be used to diagnose *N. gonorrhoeae* or *Chlamydia* using appropriate transport media as previously described. Nasopharyngeal swabs can be cultured to diagnose sinus infections, although the growth of organisms can represent infection or colonization with normal flora and is therefore not recommended [12]. Cultures must be correlated with clinical data.

Nasopharyngeal swabs are useful for diagnosis of *Bordetella pertussis* and respiratory viral infections which are identified using molecular techniques. Swabs of the anterior nares are also useful for the identification of methicillin-resistant *S. aureus* carriers, and may be used to track staphylococcal outbreaks.

Diagnosis of lower respiratory infections is made by collection of expectorated sputum, or by collection of sputum that is induced by respiratory therapy when/if the patient is unable to produce a specimen. Sputum should be collected in patients with clinical and/or radiologic signs of bacterial pneumonia. A Gram stain from the sputum can give the physician a rapid diagnosis for pneumonia. Gram stains containing an overwhelming number of squamous epithelial cells and a smaller number of WBCs likely reflect poor specimen collection. If pneumonia is still suspected, a more invasive procedure such as a bronchoscopy may need to be performed. Bronchoalveolar lavage (BAL) and/or bronchial washings can be cultured for routine as well as mycobacteria and fungi. BALs can be collected quantitatively to assess for ventilator associated pneumonia. Organisms in the quantity of $>10^4$ colonies/mL are considered significant as compared to organisms in smaller numbers, which are more likely to represent colonization. Sputum can be expectorated, without saliva or postnasal discharge, into a sterile container; induced sputa can be used to test for the presence of *Mycobacteria* spp. Sputum culture is often a critical component of diagnosis of tuberculosis. Proper identification of organisms facilitates drug susceptibility testing, which is critical in further therapeutic treatment. Contamination by bacteria and even fungi limits even the best practiced laboratory decontamination procedures. Organisms representing normal oral flora, as well as those responsible for gingivitis or oral thrush and other pharyngeal residing species, must be considered and eliminated from prognosis. It is critical to represent species collected from deep respiratory chest in the absence of poor laboratory handling. It may be difficult to collect an expectorated specimen from young children and in that case a gastric aspirate can be collected. Due to the low pH in the stomach these specimens need to be placed directly into a cup containing bicarbonate to quickly neutralize the pH to maintain viability of the organisms.

SPECIMEN COLLECTION FROM WOUNDS, ABSCESS, AND TISSUE

Pathogens that grow in superficial wound and abscess specimens can often cause skin and soft tissue infections. Rough estimates indicate there are over 1000 species of microorganism residing on skin tissue, many of which interfere with superficial wound infection identification.

Superficial wound surface should be cleaned with 70% alcohol. An aspirate of pus or a biopsy will be the most useful in obtaining the infectious organism to avoid culturing skin flora. Tissue or biopsy can be collected surgically and placed in a sterile container. Quantitative cultures may be conducted with tissues from acute wounds [13]. Organisms in a quantity of $>10^5$ are considered significant as compared to those in lower numbers which are considered more likely to be colonization. Swabs are the least useful for culture as normal flora may be picked up and not be the cause of the infection. Culturette swab specimens also do not allow for growth of fastidious organisms including anaerobes, mycobacteria, and fungi. If fluid or tissue is unable to be collected a flocked swab (Eswab) which is used for specimen collection and then broken off into a transport media is acceptable for culture of all of these organisms.

Pathogens of skin and tissue infections include *S. aureus*, beta-hemolytic streptococci, bacteria which are part of the Enterobacteriaceae group, *P. aeruginosa*, and anaerobes [13].

UNACCEPTABLE SPECIMENS

A wide range of specimens are sent to the microbiology laboratory. It is imperative that each and every specimen receive be screened to determine that they have been correctly collected and preserved. Specimens that are deemed unacceptable include those that are unlabeled or improperly labeled. Specimens received that are leaking, cracked, or broken are also rejected. Other unacceptable specimens include those that are not preserved correctly, such as those received after more than 12 hours after collection or those that are not appropriate for a particular test. Examples for improper collection tubes include blood collected in a tube with Ethylenediaminetetraacetic acid (EDTA), a lavender top tube, which is appropriate for molecular testing. The laboratory is responsible for notifying the clinical floor or subspecialty and asking for recollection of the specimen.

KEY POINTS

- Proper specimen collection is invaluable in order to isolate, identify, and further characterize infectious agents.
- Contamination can be routinely affected by anatomical site of specimen collection. Microflora of skin, as well as internal microbiota, can affect laboratory identification of clinically critical species.
- Blood cultures taken after skin decontamination with 70% alcohol followed by chlorohexidine should be filled with 5–10 mL per bottle, or, for pediatric patients volume is based on the weight of the child.
- Specimens taken from the genitourinary tract to identify sexually transmitted pathogens are usually performed by molecular methods due to the lack of sensitivity of culture for these organisms.
- UTIs are dependent on bacterial counts and disease presentation, which will ultimately determine treatment necessity.
- Stool samples are routinely tested only for *Salmonella* spp., *Shigella* spp., and *Campylobacter* spp.
- Culture of the upper respiratory tract for the diagnosis of pharyngitis is made using a throat swab or a nasopharyngeal swab.
- Culture expectorated or therapy-induced sputum is useful for the diagnosis of lower respiratory infections.
- Pathogens that grow from superficial wound and abscess specimens can often cause skin and soft tissue infections, but must be identified as a pathogen with clinical correlation.
- Specimens sent to the microbiology laboratory are screened in order to determine that they have been correctly collected and preserved. Specimens that are deemed unacceptable are incorrectly labeled, leaking, cracked, or broken samples.

REFERENCES

[1] Hall KK, Lyman JA. Updated review of blood culture contamination. Clin Microbiol Rev 2006;19:788–802.

[2] Bourbeau P, Riley J, Heiter BJ, Master R, Young C, Pierson C. Use of the BacT/Alert blood culture system for culture of sterile body fluids other than blood. J Clin Microbiol 1998;36:3273–7.

[3] Nakata MM, Lewis RP. Anaerobic bacteria in bone and joint infections. Rev Infect Dis 1984;6(Suppl 1):S165–70.
[4] von Graevenitz A, Amsterdam D. Microbiological aspects of peritonitis associated with continuous ambulatory peritoneal dialysis. Clin Microbiol Rev 1992;5:36–48.
[5] Bingen E, Lambert-Zechovsky N, Mariani-Kurkdjian P, Doit C, Aujard Y, Fournerie F, et al. Bacterial counts in cerebrospinal fluid of children with meningitis. Eur J Clin Microbiol Infect Dis 1990;9:278–81.
[6] Hussein AS, Shafran SD. Acute bacterial meningitis in adults. A 12-year review. Medicine (Baltimore) 2000;79:360–8.
[7] Meredith FT, Phillips HK, Reller LB. Clinical utility of broth cultures of cerebrospinal fluid from patients at risk for shunt infections. J Clin Microbiol 1997;35:3109–11.
[8] Larsen B, Monif GR. Understanding the bacterial flora of the female genital tract. Clin Infect Dis 2001;32:e69–77.
[9] Garcia LS, Isenberg HD. Clinical microbiology procedures handbook, vol. 1. Washington, DC: ASM Press; 2007.
[10] Kass EH. Asymptomatic infections of the urinary tract. Trans Assoc Am Physicians 1956;69:56–64.
[11] McCarter YS, Burd EM, Hall GS, Zervos M. Sharp SE, editor. Cumitech 2C: Laboratory diagnosis of urinary tract infections. Washington, DC: ASM Press; 2009.
[12] Robinson DA, Edwards KM, Waites KB, Briles DE, Crain MJ, Hollingshead SK. Clones of *Streptococcus pneumoniae* isolated from nasopharyngeal carriage and invasive disease in young children in central Tennessee. J Infect Dis 2001;183:1501–7.
[13] Bowler PG, Duerden BI, Armstrong DG. Wound microbiology and associated approaches to wound management. Clin Microbiol Rev 2001;14:244–69.

CHAPTER 3

Specific Clinical Infections

Contents

Microbiology and Molecular Diagnosis in Pathology.
DOI: http://dx.doi.org/10.1016/B978-0-12-805351-5.00003-X

INTRODUCTION

In this chapter, specific clinical infections and their diagnostic criteria are discussed. Bacteremia is associated with increased morbidity and mortality and ranks among the top seven causes of death in Europe and North America [1]. Community-acquired pneumonia is the most common infectious cause of death in the United States, but fortunately, mortality decreased from 13.5% in 1987 to 9.7% in 2005 due to improvement of antibiotic therapy [2]. Although vaccination is available for hepatitis A and B, no such vaccine is available for hepatitis C. Therefore, hepatitis C infection is a common chronic blood-borne infection in the United States and leading cause of non-AIDS mortality among human immunodeficiency virus (HIV)-infected individuals [3].

BACTEREMIA

Bacteremia is the presence of bacteria in blood. There are two clinical patterns of bacteremia, intermittent and continuous. Intermittent bacteremia is the most common pattern of bacteremia. It can occur following manipulation of or instrumentation of infected areas or in the setting of bacterial infections (such as pneumonia, arthritis, osteomyelitis, soft tissue infection, and meningitis). Continuous bacteremia is usually due to a persistent endovascular infection.

Culture of blood is considered to be the most definitive method for detection of bacteremia or fungemia. It is very important that the cultures are obtained before antimicrobial therapy is started. Circumstances in which blood cultures are especially important include known or suspected sepsis, meningitis, osteomyelitis, arthritis, endocarditis, peritonitis, pneumonia, and fever of unknown origin. Patients with bacteremia may have low quantities of bacteria in blood and bacteremia may be intermittent. It is for these reasons that it is the best to obtain multiple blood cultures and each draw should contain adequate blood volume. Prior to initiation of antimicrobial therapy, at least two sets of blood cultures taken from separate venipuncture sites should be obtained. One set consists of one aerobic and one anaerobic bottle. In certain situations, three obtaining sets may be of benefit. Such situations include when probability of bacteremia is low (e.g., pneumonia), or when the causative organism may be a common contaminant (e.g., coagulase-negative staphylococci), or when the patient has received prior antibiotic therapy. For adults, each bottle should have 5–10 mL of blood. The skin should be cleaned prior to venipuncture to avoid contamination with skin flora. Drawing blood from an indwelling intravascular catheter should be avoided, except in diagnosis of line (catheter)-related infections. Arterial blood cultures provide the same yield as venous blood cultures. Blood need not be collected at height of fever. However, best yield is before fever if fever is predictable. It is normally acceptable to collect from two separate sites within minutes of one another.

Blood Culture Bottles

Blood culture bottles have neutralizing substances for antibiotics. This is helpful in potentiating growth of organisms if the patient has received prior antibiotic therapy. Isolator systems are available for detection of filamentous and dimorphic fungi. Special broth media are available for detection of mycobacteria.

When a blood culture bottle turns positive, then a Gram stain is performed which is reported immediately. In addition, the blood is plated onto culture media. Conventional methods that are used for organism identification include physical characteristics as well biochemical profile. Relatively new technologies for organism identification include molecular methods, matrix-assisted laser desorption/ionization time-of-flight mass spectrometry (MALDI-TOF MS), and next-generation sequencing. Most significant organisms are detected within 48 hours. Fungi and fastidious organisms such as Brucella may require additional time. Blood culture bottles are routinely held for 5 days.

Common Contaminants of Blood Culture

- *Propionibacterium acnes*
- *Corynebacterium species*
- *Bacillus species*
- *Coagulase-negative staphylococci*
- *Viridans streptococci*

Contamination should be suspected when bacterial growth first occurs in blood cultures after 72 hours of incubation or in a single bottle out of multiple sets. However, certain fastidious organisms require additional time for growth (e.g., HACEK (*Haemophilus* species, *Aggregatibacter* species, *Cardiobacterium hominis*, *Ekinella corrodens* and *Kingella* species) organisms). Individuals who have received antimicrobial therapy may also have bacteria in their blood which requires additional duration of therapy.

INFECTIVE ENDOCARDITIS

Infective endocarditis (IE) refers to infection of cardiovascular structures including one or more heart valves, atrial or ventricular endocardium, and intracardiac devices. Infective carditis is more common in developing countries. Without appropriate treatment mortality approaches 100%. Endocarditis is seen in situations where there is presence of organisms in the blood and there is abnormal cardiac endothelium which facilitates their adherence and subsequent growth.

Risk factors for developing IE include:

- Age: Individuals over the age of 60 years are at increased risk.
- Male are more prone than women.
- IV drug abuse.
- Poor dentition, dental infection, and dental procedures that involve manipulation of gingival tissue (not routine dental cleaning).
- Structural heart disease: For example, valvular disease (including mitral valve prolapse), congenital heart disease, and prosthetic heart valves.
- Presence of intravascular device.
- Chronic hemodialysis.
- HIV infection.

Organisms Implicated in Infective Endocarditis

Several organisms cause IE. These organisms include:

- *Staphylococcus aureus* (this is responsible for about one in three cases).
- Viridans group streptococci (second commonest organism).
- Enterococci.
- Coagulase-negative staphylococci.
- HACEK organisms.
- Rarely, Gram-negative organisms.
- Yeast.

Staphylococcal IE is a common cause of healthcare-related IE as well as community-acquired IE.

Clinical Features

Clinical features are variable and at times may be vague and nonspecific. A high index of suspicion is valuable for early diagnosis. Clinical features include:

- Fever (most common symptom).
- *Nonspecific symptoms*: Malaise, fatigue, headache, and weight loss.
- Cardiac murmurs.
- Splenomegaly.
- Septic emboli (e.g., stroke, septic arthritis, and renal infarction).
- Glomerulonephritis.
- Splinter hemorrhages (nonblanching, linear reddish-brown lesions under the nail bed).
- Janeway lesions: Nontender erythematous macules on the palms and soles.
- Osler nodes: Tender subcutaneous purplish nodules mostly on the pads of the fingers and toes.
- Roth spots: Hemorrhagic lesions of the retina with pale centers.

A modified Duke criterion is used to make a clinical diagnosis of endocarditis [4].

Investigations

- *Blood cultures*: It is recommended that two-three sets of blood cultures should be obtained from separate venipuncture sites prior to initiation of antibiotic therapy. As most patients with IE have continuous bacteremia, it is acceptable for blood cultures to be collected at any time. It is not necessary for them to be obtained at the time of fever or chills. Most clinically significant bacteremias are detected within 24–48 hours; common and fastidious pathogens (such as members of the HACEK group) may be detected within 5 days of incubation. The optimal volume of blood for each blood culture in adults is 10 mL. In patients who have received recent antimicrobial therapy, additional blood cultures are recommended.

  Follow-up blood cultures should be obtained 48–72 hours after antimicrobial therapy is begun and repeated every 48–72 hours until clearance of bacteremia.
- *Echocardiography*: Culture-negative endocarditis is defined as endocarditis with no definitive microbiologic etiology following inoculation of at least three independently obtained blood samples in a standard blood culture system, with negative cultures after 5 days of incubation.

Culture-negative IE should also be suspected in patients with vegetation on echocardiogram and no clear microbiologic diagnosis. Some causes of culture-negative IE may be identified via serology or polymerase chain reaction (PCR); these include *Coxiella burnetii*, *Bartonella* spp., *Chlamydia* spp., *Legionella* spp., *Mycoplasma*, and *Brucella*.

Antibiotic Therapy

In general, therapy for IE is dictated by the organism isolated from blood cultures; cultures are positive in over 90% of patients with such disease. For patients with suspected IE who present without acute symptoms, empiric therapy is not always necessary, and therapy can await blood culture results.

For acutely ill patients with signs and symptoms strongly suggestive of IE, empiric therapy may be necessary. Such empiric therapy should be administered after at least two sets of blood cultures have been obtained from separate venipunctures and ideally spaced over 30–60 minutes. The choice of empiric therapy should take into consideration the most likely pathogens. In general, empiric therapy should cover staphylococci (methicillin susceptible and resistant), streptococci, and enterococci. Vancomycin is an appropriate choice for initial therapy in most patients.

The duration of therapy depends on the pathogen and site of valvular infection; it must be sufficient to eradicate microorganisms growing within the valvular vegetations. The duration of therapy should be counted from the first day of negative blood cultures (for cases in which blood cultures were initially positive). Most patients are treated parenterally with regimens given for up to 6 weeks. In general, 6 weeks of treatment is appropriate for patients with virulent or relatively resistant pathogens, secondary cardiac or extracardiac complications, and in the setting of prolonged infection prior to diagnosis. Shorter therapeutic regimens may be effective in selected patients with right-sided endocarditis and with endocarditis due to highly susceptible *V. streptococci* treated with synergistic antimicrobials.

JOINT INFECTIONS

Joints may be infected with bacteria, virus, fungus, or mycobacteria. Infection by bacteria can lead to rapid joint destruction. Bacterial infection is most often a result of hematogenous spread. However, it can also result from a bite or other trauma, direct inoculation of bacteria during joint surgery, or, in rare cases, following extension of preexisting bony infection through the cortex into the joint space. If a patient has bacteremia, organisms are more likely to localize in a joint with preexisting arthritis. Common predisposing factors for hematogenous dissemination include IV drug use, alcoholism, presence of indwelling catheters, and immunodeficiency states. Neonates and older adults are at highest risk.

Many pathogens may cause bacterial arthritis and is monomicrobial. *S. aureus* is the most common bacterium infecting adult joints, followed by streptococci. Septic arthritis due to Gram-negative bacilli may be seen in the setting of trauma, intravenous drug users, neonates, older adults, and in association with underlying immunosuppression. In sickle cell patients infection with *Salmonella* may be seen. In neonates *Kingella* associated joint infections may be seen with a predilection for involvement of ankle joints.

P. acnes are relatively common cause of prosthetic joint infection following shoulder arthroplasty but is a rare cause of infection following hip or knee arthroplasty.

Prosthetic Joint Infections

The rate of prosthetic joint infection ranges between 0.5% and 2% for hip, knee, and shoulder replacements. Clinically there are three types of infection, early, delayed, and late.

Early infections: this takes place within 3 months of surgery and the infection is acquired during implantation. Infecting organisms are typically virulent.

Delayed infections: this occurs between 3 and 12 months postsurgery. The infection is acquired during implantation but less virulent organisms are implicated. Examples include *Propionibacterium* species, coagulase-negative staphylococci, or enterococci.

Late infections: these occur beyond 12 months of surgery. Infection is by hematogenous spread from another site.

Clinical features include single, painful large joint (knee is the joint most often involved).

Diagnosis

- Joint fluid analysis: Gram stain, culture, cell count with differential, and analysis for crystals.
- Blood culture.

Treatment

Treatment of acute bacterial arthritis requires antibiotic therapy and joint drainage. For Gram-positive cocci vancomycin is an appropriate antibiotic. For Gram-negative rods a third-generation cephalosporin is appropriate.

Examples of specific type of bacterial arthritis:

- *Meningococcal arthritis*: this may complicate meningococcemia and may present as migratory polyarthritis. The joints cultures are typically negative.
- *Gonococcal arthritis*: most common cause of septic arthritis in previously fit young adults. Culture is usually positive for gonococcus from the genital tract; joints cultures tend to be negative.
- *IE*: patients with IE are well known to develop arthritis.
- *Lyme disease*: patient with Lyme disease due to *Borrelia burgdorferi* may develop oligoarthritis and a typical skin rash known as erythema chronicum migrans.
- *Brucellosis*: *Brucella melitensis* is the leading cause of brucellosis and associated arthritis.
- *Tuberculous arthritis*: about 1% of patient with TB develop arthritis. In children tuberculous arthritis may develop as the primary disease. In adults it is most often due to spread from the primary focus. The hip, knee, and spine are common areas of involvement.

RESPIRATORY TRACT INFECTION

Pneumonia is a common respiratory tract infection which is defined as inflammation of lung parenchyma. Pneumonia is one of the leading causes of death in the United States. The lungs are constantly exposed to microorganisms from the air we breathe and aspiration of nasopharyngeal flora. However, there are numerous defense mechanisms:

- Mucociliary apparatus.
- Cough reflex and gag reflex.
- Local immunoglobulin production.
- Polymorphonuclear neutrophils (PMNs), macrophages, and cell-mediated immunity.

These defense mechanisms may be compromised:

- Damage to the mucociliary apparatus by smoke, toxins, viral infections, or due to immotile cilia syndrome.
- Suppressed cough reflex due to anesthesia, drugs, alcohol, or in unconscious patients.

- Decreased immunoglobulin production: hypogammaglobulinemia.
- Neutropenia from chemotherapy and steroids.

Pneumonia may be classified anatomically or by etiology.

Anatomical:

- Lobar pneumonia (pneumonia affecting one or more lobe).
- Lobular pneumonia (pneumonia affecting lobules of lung, a.k.a. bronchopneumonia).
- Interstitial (interstitial pneumonia refers to pneumonia where the inflammation is confined to the walls of the alveoli. This is typically seen in viral pneumonia and atypical pneumonias).

Etiological:

- Bacterial.
- Tuberculous.
- Viral.
- Fungal.

The following terminology is also widely used for pneumonia:

- Community-acquired pneumonia.
- Atypical pneumonia.
- Nosocomial pneumonia.
- Aspiration pneumonia.
- Chronic pneumonia.
- Pneumonia in the immunocompromised host.

Community-Acquired Pneumonia

Community-acquired pneumonia is a significant cause of morbidity and mortality in adults. The overall annual incidence of community-acquired pneumonia ranges from 5 to 11 per 1000 persons, with more cases occurring in winter months [5]. Important bacterial causes of pneumonia include:

- *Streptococcus pneumoniae*: predisposing causes include: (1) chronic diseases, congestive heart failure (CHF), chronic obstructive pulmonary disease (COPD), and diabetes, (2) immunodeficiency, and (3) absent spleen.
- *Haemophilus influenzae*: predisposing causes include (1) COPD and (2) cystic fibrosis.
- *Moraxella catarrhalis*: Predisposing cause include COPD.
- *S. aureus*: Predisposing causes include: (1) viral respiratory illness (e.g., measles and influenza) and (2) IV drug abuse. It is an important causative agent for nosocomial pneumonia and has high incidence of complications.

- *Klebsiella pneumoniae*: Predisposing causes include (1) alcoholics and (2) debilitated and malnourished states.

These and others typically are the causes of community-acquired pneumonia and result in lobar or bronchopneumonia.

Atypical Pneumonia

In this type of pneumonia, classical features of pneumonia may not be seen. The patient lack features of consolidation as inflammation is mainly interstitial. The patient has more symptoms than signs. Leukocytosis may not be marked. Standard antibiotics used in community-acquired pneumonia will not be effective. Causes of atypical pneumonia include:

- *Mycoplasma pneumoniae* (in about 50% of cases there is presence of cold agglutinins).
- *Chlamydia pneumoniae.*
- *C. burnetii* (Q fever).
- *Viruses*: Influenza A and B, respiratory syncytial virus (RSV), adenovirus, rhinovirus, herpes simplex virus (HSV), varicella zoster virus (VZV), cytomegalovirus (CMV). Influenza virus is the most common viral pneumonia in the nonimmunocompromised host. May be complicated by Staphylococcal pneumonia. Adenovirus pneumonia is typically mild; may be severe in the immunocompromised host. RSV is the most common cause of serious respiratory infection in infants. CMV pneumonia is mostly seen in immunocompromised host (transplant patients and AIDS).

Nosocomial Pneumonia

This is seen in hospitalized and very ill patients who may be immunocompromised patients, patients on prolonged antibiotic therapy, and ventilator associated pneumonia. Often the implicating organisms are Gram-negative bacteria, including *Pseudomonas* and *S. aureus*. *Pseudomonas* invades blood vessels, resulting in necrosis of blood vessels as well bacteremia, with a high mortality.

Aspiration Pneumonia

This type of pneumonia occurs due to aspiration of gastric contents, most often into the right lung. This is most often seen in patients with loss of gag reflex which may occur in unconscious patients, stroke patients, and individuals with neuromuscular disorders (e.g., motor neuron disease, Guillain Barre Syndrome, etc.). Implicated organisms are Gram-negative, anaerobes, *Pseudomonas*, and *S. aureus*.

Chronic Pneumonias

This group includes tuberculosis (TB) (defined as chronic granulomatous disease due to *Mycobacterium tuberculosis*) and fungal infections. In the immunocompetent the lesion is localized. In the immunocompromised there may be systemic disease.

Individuals who are at risk for TB include:

- Human immunodeficiency virus (HIV).
- Diabetes mellitus.
- Chronic lung disease.
- Chronic renal failure.
- Malnutrition.
- Alcoholism.
- Immunosuppression.
- Hodgkin lymphoma.

In pulmonary TB, organism gains entry through inhalation (from another individual with active TB). Bacteria gain entry into macrophages (mediated by receptors such as mannose receptor). However, bacteria are not killed but may proliferate. Macrophages with bacteria disseminate and seed in different parts of the body. After about 3 weeks, cell-mediated immunity takes place. Mycobacterial antigens are presented to CD4+ T cells by macrophages, TH1 cells are generated, interferon (IFN)-gamma is released, and macrophages are activated. This leads to recruitment of monocytes, and macrophages are converted to epithelioid histiocytes. Granuloma formation takes place. TB bacilli settle in lower part of upper lobe or upper part of lower lobe. Lesion described above takes place. This is referred to as Ghon focus. Bacilli drain to regional lymph node with similar changes. This with Ghon focus is called Ghon complex. The bacilli are walled off and remain dormant. Patient is now sensitized and tuberculin skin test will be positive. If patient is immunocompromised then progressive primary TB ensues. This is like bacterial pneumonia. Dissemination with resultant TB meningitis and miliary TB may ensue.

In immunocompetent individual's bacteria remains dormant waiting for reactivation. Reactivation results in postprimary TB (or secondary TB). Less than 5% of individuals with primary TB develop secondary TB. The location of secondary TB is classically at apex of lobes and cavitation is a key feature. When lesion erodes into bronchus, sputum becomes contagious and is acid fast bacillus (AFB) positive. Erosion in to blood vessel will result in hemoptysis. Disease can spread to different parts of the lungs through airways: progressive pulmonary TB. With hematogenous spread bacilli may disseminate to different parts of the body: Miliary TB. Organs affected include liver, spleen, bone marrow, kidneys, adrenals, and meninges.

In HIV positive individual's manifestation of TB depends largely on the CD4 cell count. If CD4+ counts >300 cells/mm^3, usual secondary TB ensues. If the CD4+ count is <200, then there is more likely development of progressive pulmonary TB (noncavitary consolidation). In such cases the sputum may be negative for AFB. Tuberculin skin test will be negative due to immune anergy and lesions may lack granulomas. Cervical lymphadenitis and intestinal TB are two examples of extrapulmonary TB.

Interferon-Gamma Release Assays for Diagnosis of Latent Tuberculosis Infection

The development of interferon-gamma release assays (IGRAs) is used in the diagnosis of latent tuberculosis infection (LTBI). IGRAs are in vitro blood tests of cell-mediated immune response; they measure release of interferon (IFN)-gamma by T cells after stimulation by antigens unique to *M. tuberculosis*. Antigens used include early secreted antigenic target 6 and culture filtrate protein 10. The number of IFN-gamma producing T cells or the amount of IFN-gamma may be quantified as results.

The goal of testing for LTBI is to identify individuals who are at increased risk for the development of TB and therefore who would benefit from treatment of LTBI. IGRAs cannot distinguish between latent infection and active TB. IGRAs are not affected by prior BCG administration.

Two IGRAs are commercially available: The QuantiFERON-TB Gold In-Tube (QFT-GIT) assay and the T-SPOT.TB assay. The QFT-GIT assay is an enzyme-linked immunosorbent assay-based, whole-blood test that measures IFN-gamma produced in response to TB antigens. The T-SPOT.TB is an enzyme-linked immunospot assay which measures IFN-gamma producing T cells in response to TB antigens.

IGRAs have specificity >95% for diagnosis of LTBI.

IGRA sensitivity is diminished by HIV infection.

Fungal Chronic Pneumonias

Histoplasmosis, Coccidioidomycosis, and Blastomycosis are typically localized disease in immunocompetent individuals and disseminated disease in immunocompromised state. These are dimorphic fungi that exist in the mold form in soil and in humans they are in the yeast form. In culture, the mold form is identified. In tissue or body fluids, the yeast form is detected.

Upon initial exposure "flu" like symptoms may develop with histology similar to primary TB. In the vulnerable host, a cavitary lesion similar to secondary TB is seen. In immunocompromised host disseminated disease little granuloma formation takes place.

Coccidioidomycosis, also known as San Joaquin Valley Fever, is due to *Coccidioides*, an infectious fungal disease confined to the western hemisphere and endemic in American deserts. *Histoplasma capsulatum* is found throughout the world. It is endemic in certain areas of the United States, particularly in states bordering the Ohio River valley and the lower Mississippi River. *Blastomyces dermatitidis* is a dimorphic fungal pathogen, found primarily in the Mid-West and Northern United States and Canada. It exists in the soil in a filamentous form that produces spores. The natural reservoir seems to be associated with rivers and lakes. Blastomyces is endemic to the Mississippi and Ohio River valleys and the vicinity of the Great Lakes.

Pneumonia in Immunocompromised People

In the immunocompromised people in addition to the usual organisms which are implicated in the immunocompetent individuals, other organisms which may be involved are:

- CMV, VZV and HSV (herpes simplex virus).
- *Pneumocystis jiroveci.*
- Mycobacterium avium-intracellulare.
- Invasive aspergillosis.
- Cryptococcosis.
- Mucormycosis.

Pneumocystis Jiroveci Pneumonia

This organism was previously known as *Pneumocystis carinii* and was thought to be a protozoan. It is now considered to be a fungus. Most humans are exposed to it and the organism remains dormant. Reactivation occurs if the host is immunocompromised, especially in AIDS. A CD4 count of 200 or below ($/mm^3$) is a strong predictor of *P. jiroveci* pneumonia.

URINARY TRACT INFECTION

Urinary tract infection (UTI) is a common infection for females. UTI may take the form of asymptomatic bacteria, urethritis, cystitis, and pyelonephritis. More than 85% of cases of UTI are due to organisms that are normally found in the gastrointestinal tract. The process starts with urethritis and with ascending infection cystitis may follow. Ascending infection from the bladder may result in pyelonephritis. Hematogenous route of infection is also well described as a cause of pyelonephritis, however it is less common. Urinary tract obstruction, vesicoureteric reflux, and intrarenal reflux all aid in the process of ascending infection.

Escherichia coli is the most common bacterial cause of UTI; it accounts for approximately 80% of UTI in children. Other Gram-negative bacterial pathogens include *Klebsiella*, *Proteus*, *Enterobacter*, and *Citrobacter*. Gram-positive bacterial pathogens include *Staphylococcus saprophyticus*, *Enterococcus*, and, rarely, *S. aureus*. Mycobacterial and fungal organisms may also be implicated. In immunocompromised patients' viruses such as polyoma virus, CMV, and adenoviruses may also cause UTI. Acute pyelonephritis is a serious complication of UTI and repeated attacks may lead to chronic pyelonephritis.

Predisposing features of acute pyelonephritis include:

- Urinary tract obstruction.
- Urinary tract catheterization.
- Diabetes mellitus.
- Immunosuppression.
- Vesicoureteral reflux.

Clinical features include:

- Frequency of micturition.
- Dysuria.
- Hematuria.
- Suprapubic pain and tenderness.

Diagnosis: Urine dipstick test for UTI; individuals with UTI may demonstrate the following positive dipsticks:

- Hematuria.
- Positive leukocyte esterase.
- Positive nitrite.

Urine microscopy; pus cells, and RBCs may be seen.

Urine culture: it has been recommended that routine microbiological processing of urine specimens include quantitative plating onto blood agar medium along with a selective and differential agar such as MacConkey agar for Gram-negative organisms.

Criteria for Diagnosis

Symptomatic young women: $\geq 10^2$ coliform organisms/mL of urine or $\geq 10^5$ (clean catch) any pathogen/mL of urine or $>10^4$ organisms from catheterized urine or any growth of pathogens from urine obtained by suprapubic puncture.

Symptomatic men: $\geq 10^3$ any pathogen/mL of urine.

Asymptomatic patients: $\geq 10^5$ any pathogen/mL of urine on two occasions.

Treatment

Approximately 50% of *E. coli* are resistant to amoxicillin or ampicillin. In addition, increasing rates of *E. coli* resistance to first-generation cephalosporins and trimethoprim-sulfamethoxazole have also been reported. Quinolones, nitrofurantoin, and oral second generation cephalosporins such as cefuroxime are appropriate agents for treatment of UTI. Third-generation cephalosporins and aminoglycosides are appropriate first-line agents for empiric treatment of UTI in children.

VIRAL HEPATITIS

Common causes of viral hepatitis include infection with:

- Hepatitis A.
- Hepatitis B.
- Hepatitis C.
- Hepatitis D.
- Hepatitis E.

Other causes of hepatitis due to viral infections include:

- Epstein-Barr virus.
- CMV.
- Herpes simplex virus.
- HIV.
- Adenovirus.
- Yellow fever virus.

Although no definite and consistent seasonal pattern has been observed with acute hepatitis viral infection, there are some indication of spring and summer peak for hepatitis A, B, C, and E infection [6]. Key points of five hepatitis virus are summarized in Table 3.1.

Hepatitis A

Viral hepatitis may be caused by the hepatitis A virus (HAV) and occurs worldwide either sporadically or as an epidemic. It is usually a self-limited illness that does not become chronic hepatitis. Fulminant hepatic failure develops in less than 1% of patients with hepatitis A. Hepatitis A infection

Table 3.1 Key points of five hepatitis virus

	A	B	C	D	E
Hepatitis virus type	Hepatitis A is a Picornavirus; RNA virus; 27 nm	Hepatitis B is a Hepadna virus; DNA virus; 42 nm	Hepatitis C is a Flavivirus; RNA virus; 50 nm	Hepatitis D is a Delta virus; RNA virus; 36 nm	Hepatitis E is a Hepevirus; RNA virus; 27 nm
Transmission	Mainly fecal-oral	IV; blood and blood products as well as sexual and vertical	Blood and blood products; occasionally sexual and vertical	Mainly blood and blood products	Mainly fecal-oral
Incubation period	2–3 weeks	1–5 months	Weeks to months	Months	Weeks
Mortality from acute hepatitis	<0.5%	<1%	<1%		1%–2%; in pregnant women as high as 10%–20%
Carrier state	No	Yes	?	Yes	No
Chronic liver disease	No	Yes	Yes	Yes	No
Hepatocellular cancer	No	Yes	Yes	Yes	No
Vaccination	Available	Available	No	No	No

confer lifelong immunity, and infection is preventable via vaccination. HAV infection is usually transmitted by the fecal-oral. Blood-borne transmission can occur but is uncommon. Hepatic injury occurs as a result of the host immune response to HAV. Viral replication occurs in the hepatocyte cytoplasm; hepatocellular damage and destruction of infected hepatocytes by CD8+ T lymphocytes and natural killer cells. An excessive host response is associated with severe hepatitis.

Clinical features include:

- Initial period of nausea, vomiting, fatigue, fever, and malaise.
- Followed by jaundice and hepatomegaly.
- Abdominal pain.
- Splenomegaly.
- Skin rash.
- Arthralgia.

Laboratory findings include:

- Abnormal liver function tests (elevated liver enzymes and elevated bilirubin).
- Elevated IgM, anti-HAV antibodies.

Serum IgM antibodies are detectable at the time of symptom onset, peak during the acute or early convalescent phase of the disease, and remain detectable for approximately 3–6 months. Serum IgG antibodies, once they appear remain for decades and provide immunity.

Hepatitis B

Hepatitis B is present worldwide. The carrier rate is lower in the United States when compared to Asian and African countries. In the United States the carrier rate is around 1%–2%, whereas in Asia and African countries it can be as high as 10%–20%.

The hepatitis B virus (HBV) consist of a nucleocapsid (27 nm), surrounded by an outer surface protein (HBsAg). This HBsAg is produced as excess and can be found independently in serum and body fluids. HBsAg has several antigenic determinants. These are a, d, y, w, and r. Combinations of these determinants are used to classify HBV genotypes. There are eight major genotypes described, A–H. However, I and J have also been described. Genotype A is seen mainly in North America, North West Europe, and Central Africa. Type C are seen mainly in South East Asia; D is seen in Southern Europe, India, and Middle East; E in West Africa; F in South and Central America; G in France and United States; and H in Central and South America. Certain genotypes respond better to IFN treatment (e.g., A > B and C > D).

Tests for Hepatitis B:

HBsAg: Appears in blood in about 1–10 weeks after infection and then disappears; if HBsAg persists then this indicates a carrier state or chronic infection. HBeAg: Becomes present with acute infection and then disappears; if persists implies persistent viral replication and patients is infectious. HBcAg: Not tested.

Anti-HBc: First antibody to appear; may be the only marker for hepatitis infection when HBsAg is negative and anti-HBs has not appeared.

Anti-HBe: Second antibody to appear; implies decreased infectivity.

Anti-HBs: Last antibody to appear; implies immunity.

HBV DNA: Found in serum and liver; implies viral replication.

Hepatitis C

Hepatitis C virus (HCV) can cause acute hepatitis or chronic liver disease. Six different genotypes are identified. Type 1a and 1b account for 70% of cases in the United States.

Most (90%) acute infections are asymptomatic. About 90% of asymptomatic patients may develop chronic liver disease in the future. If the acute infection is symptomatic a higher percentage of these patients clear the virus. About 50%–75% of this group may develop chronic liver disease.

Tests for HCV:

Anti-HCV, IgG, and IgM.

HCV-RNA by PCR.

Hepatitis D

Hepatitis D Virus (HDV) is incomplete RNA particle enclosed in a shell of HBsAg. Hepatitis D infection can occur as co-infection with hepatitis B or as superinfection in an individual already infected with hepatitis B. Hepatitis D essentially makes outcome of hepatitis B worse.

Tests for hepatitis D usually consists of Anti-HDV.

Hepatitis E

Hepatitis E virus (HEV) is an RNA virus which causes hepatitis similar to hepatitis due to A. It is enterally transmitted and does not cause chronic liver disease.

Tests for hepatitis E:

Anti-HEV, IgG, and IgM.

GASTROINTESTINAL INFECTIONS

Diarrheal diseases represent one of the five leading causes of death worldwide. The vast majority of cases of acute diarrhea in adults are of infectious etiology. Most cases of acute infectious diarrhea are viral. Diarrhea may be defined as the passage of loose or watery stools, typically at least three times in a 24-hour period. Diarrhea may be acute, persistent, or chronic. If the duration of diarrhea is up to 2 weeks, then this is acute diarrhea. If the duration is greater than 2 weeks but less than a month, it is considered to be persistent diarrhea. Diarrhea lasting more than a month is chronic diarrhea.

Invasive diarrhea, or dysentery, is defined as diarrhea with visible blood or mucus, in contrast to watery diarrhea. Dysentery is commonly associated with fever and abdominal pain.

Established pathogens of the gastrointestinal tract infection include bacteria, viruses, parasites and fungus.

Commonly encountered bacteria causing gastrointestinal tract infections include:

- *Clostridium difficile.*
- *Clostridium perfringens.*
- *Aeromonas* spp.
- *Plesiomonas shigelloides.*
- *Yersinia enterocolitica.*
- *Campylobacter jejuni.*
- *Salmonella.*
- *Shigella.*
- *Vibrio cholerae.*
- *Vibrio parahaemolyticus.*
- Enterotoxigenic *E. coli.*
- Enteropathogenic *E. coli.*
- Enterohemorrhagic *E. coli* (EHEC).
- Enteroinvasive *E. coli.*
- Enteroaggregative *E. coli.*

Commonly encountered viruses causing gastrointestinal tract infections include:

- Astrovirus ssRNA.
- Hepatitis A ssRNA.
- Norwalk virus.
- Rotavirus dsRNA.

Commonly encountered parasites causing gastrointestinal tract infections include:

- *Cryptosporidium* spp.
- *Entamoeba histolytica.*
- *Echinococcus* spp.
- *Giardia lamblia.*
- *Dientamoeba fragilis.*
- *Balantidium coli.*
- *Chilomastix mesnili.*
- *Ascaris lumbricoides.*
- *Enterobius vermicularis.*
- *Trichuris trichiura.*
- *Ancylostoma duodenale.*
- *Nectar americanus.*
- *Strongyloides.*
- *Taenia saginata.*
- *Taenia solium.*
- *Diphyllobothrium latum.*
- *Schistosoma* spp.
- *Fasciola hepatica.*
- *Clonorchis sinensis.*
- *Paragonimus westermani.*

Most cases of acute diarrhea are mild or transient and do not require work up. However, individuals with bloody diarrhea, fever, and significant abdominal pain should be investigated.

Causes of visible bloody diarrhea:

- EHEC.
- *Campylobacter.*
- *Salmonella.*
- *Shigella.*

At the time of initial evaluation of an individual with acute diarrhea, specific exposure/history may be sought:

- Unpasteurized dairy products, raw, or undercooked meat or fish.
- Exposure to animals (poultry, turtles, and petting zoos) has been associated with Salmonella infection.
- Recent antibiotic use (for risk of *C. difficile*).
- Travel history.

Testings for diagnosis of gastrointestinal tract infection include:

- Stool culture.
- Testing for *C. difficile.*
- Testing for shiga toxin.
- Testing for fecal leukocytes.
- Testing for amebiasis.

Campylobacter Infections

Campylobacter infection is an important cause of acute diarrhea. *Campylobacter* enteritis is typically caused by *C. jejuni* or *Campylobacter coli.* The clinical features of *Campylobacter* enteritis due to *C. jejuni* and *C. coli* are indistinguishable from one another and from illness due to other bacterial pathogens, such as *Salmonella* or *Shigella.*

Transmission of *Campylobacter* infection is by ingestion of contaminated animal products such as raw milk, uncooked chicken, and contaminated water. Campylobacter is sensitive to gastric acid. Conditions or medications which reduce gastric acid production will predispose an individual to infection.

Campylobacter infections may mimic clinical symptoms of appendicitis. There may be an association between *C. jejuni* infections and lymphomas. Campylobacter infections are associated with reactive arthritis and Guillain Barre Syndrome.

Salmonella Infections

Salmonellae cause a broad range of infections, including gastroenteritis, enteric fever, and focal infections such as osteomyelitis. Salmonella may cause typhoid or enteric fever by infections with *Salmonella typhi* or *S. paratyphi.*

Shigella Infections

Shigella species can cause of bacterial diarrhea worldwide. *Shigella* organisms can survive transit through the stomach since they are less susceptible to acid than other bacteria; for this reason as few as 10–100 organisms can cause disease.

The different types of *Shigella* include:

- *Shigella flexneri.*
- *Shigella sonnei.*
- *Shigella dysenteriae.*
- *Shigella boydii.*

Shigella infections may also be responsible for:

- Reactive arthritis (reactive arthritis has also been described following infection with *C. jejuni*, *Salmonella enteritidis*, *Shigella typhimurium*, *Y. enterocolitica*, and *Yersinia pseudotuberculosis*).
- Hemolytic-uremic syndrome (it is most often due to infection with enterohemorrhagic *E. coli* (particularly type O157:H7), but may also be due to *S. dysenteriae*).

Clostridium difficile Infections

C. difficile is a spore-forming, toxin-producing, gram-positive anaerobic bacterium that causes antibiotic-associated colitis. It colonizes the intestinal tract after the normal gut flora has been altered by antibiotic therapy. *C. difficile* infection (CDI) is one of the most common healthcare-associated infections and a significant cause of morbidity and mortality among older adult hospitalized patients. CDI can vary from an asymptomatic carriage to fulminant disease with toxic megacolon.

Asymptomatic *C. difficile* carriage occurs in about 20% of hospitalized adults; these patients shed *C. difficile* in stool but do not have diarrhea or other clinical symptoms. The asymptomatic carriage rate may be as high as 50% in long-term care facilities.

Watery diarrhea with or without abdominal pain is the cardinal symptom of *C. difficile*-associated diarrhea (CDAD) with colitis (≥3 loose stools in 24 hours). Symptoms of CDI typically occur in the setting of antibiotic therapy; they may begin during antibiotic therapy or 5–10 days following antibiotic therapy. Rarely, symptoms present as late as 10 weeks after antibiotic therapy has been stopped. The antibiotics most frequently implicated in predisposition to CDI are:

- Fluoroquinolones.
- Clindamycin.
- Cephalosporins.
- Penicillins.

Up to 25% of patients experience recurrent *C. difficile* within 30 days of treatment.

The diagnosis of CDI should be suspected in patients with clinically significant diarrhea (≥3 loose stools in 24 hours) or ileus in the setting of relevant risk factors (including recent antibiotic use, hospitalization, and advanced age). The diagnosis is established via a positive laboratory stool test for *C. difficile* toxins or *C. difficile* toxin gene.

For patients with diarrhea and suspected CDI, liquid stool should be sent for *C. difficile* testing. Laboratory testing does not distinguish between CDAD and asymptomatic carriage. If patients are asymptomatic specific treatment is not required.

Findings of pseudomembranous colitis (severe inflammation of the inner lining of the bowel) on radiographic or endoscopic examination are highly suggestive of CDI and should prompt laboratory testing.

There is no benefit for repeat laboratory testing or testing for cure. There is also no role for laboratory testing in asymptomatic patients or in patients receiving treatment for acute CDI; stool assays may remain positive during or after clinical recovery. Laboratory diagnosis of CDI requires demonstration of *C. difficile* toxin(s) or detection of toxigenic *C. difficile* organism(s).

Viruses

Astrovirus, HAV, and the Norwalk virus are single stranded RNA viruses and are all implicated in causing acute gastroenteritis. HAV is also responsible for causing acute hepatitis. The rotavirus is a double stranded RNA virus which causes severe diarrhea in infants and young children.

Parasites: See chapter on parasitology (Chapter 10: Infections Caused by Parasites).

CENTRAL NERVOUS SYSTEM (CNS) INFECTIONS

Meningitis is an inflammatory disease of the leptomeninges and is characterized by increased number of white blood cells in the cerebrospinal fluid (CSF). It is one of the top 10 most common infectious causes of death. Individuals who survive may have residual neurologic deficit. Bacterial meningitis can be community acquired or healthcare associated. The common causes of community-acquired bacterial meningitis in adults are *S. pneumoniae* and *Neisseria meningitidis*. In individuals over 50 years and with cell-mediated immunodeficiency *Listeria monocytogenes* is an important cause.

In neonates, group B streptococcus (GBS), *Listeria* and *E. coli* are the leading causes. In young children, *H. influenza* is a well-recognized etiologic agent in children between 5 months and 5 years. The usual causes of healthcare-associated bacterial meningitis or meningitis in patients with artificial devices such as shunts are staphylococci and aerobic Gram-negative rods.

Neisseria meningitidis

This bacteria can cause:

- Bacteremia without sepsis.
- Meningococcemia without meningitis.
- Meningitis with or without meningococcemia.
- Meningoencephalitis.
- Chronic meningococcemia.

Neisseria meningitidis infections may cause petechiae and purpura. Arthritis is also another feature associated with such infections.

Streptococcus pneumoniae

This is the most common etiologic agent for bacterial meningitis in adults. Human nasopharynx is the main reservoir of this organism and serves as a source of person to person transmission. Pathogenesis is most likely due to direct extension from sinuses or middle ear or as a result of bacteremia.

Listeria

Meningitis due to *Listeria* may be seen in neonates as well as elderly patients. *Listeria* infection may be associated with consumption of non-pasteurized milk. Clinically the condition may mimic meningitis by group B streptococci. In contrast to streptococci *Listeria* is resistant to cephalosporins.

Group B Streptococcus (GBS)

About 10%–30% of pregnant women are colonized by GBS in their vagina or rectum. Organisms may be found in amniotic fluid and neonates may become infected by aspiration during birth.

Clinical features of meningitis include:

- Fever.
- Neck rigidity.
- Altered mental status.
- Headache.

In addition some may develop seizures and focal neurologic deficit.

Relevant investigations:

- Complete blood count: May show leukocytosis.
- Blood cultures: May be positive.
- Lumbar puncture: Increased opening pressure, increased WBC count, high protein, low glucose, positive cultures; Gram stain may be positive

(Gram stain has a sensitivity of 60%–90% and a specificity of 97%–100%. PCR studies may also be ordered from CSF specimens.

Meningitis and or encephalitis may be aseptic and this is when routine bacterial cultures are negative. Etiologic agents in these cases include:

- Enteroviruses, arboviruses (togaviruses, flavivirus, West Nile virus, JC virus), HSV, HIV, VZV, mumps, rubella, lymphocytic choriomeningitis virus, and post-measles encephalitis.
- Mycobacteria.
- Weakly acid-fast bacteria.
- Fungi (e.g., cryptococci and coccidioidal infections).
- Spirochetes (*Treponema pallidum* and *B. burgdorferi*).
- Ameba (Naegleria, Acanthamoeba, and Balamuthia).
- Parameningeal infections.
- Medications.
- Malignancies.

Cryptococcus spp.

Cryptococcus is round budding yeasts without pseudohyphae. It has capsules which may be stained with India in or mucicarmine. *Cryptococcus neoformans var neoformans* causes meningitis in HIV patients and *C. neoformans var gattii* in non-HIV patients.

Many cases of aseptic meningitis follow a self-limiting course and do not require specific therapy.

Encephalitis

Individuals with meningitis do not have abnormal cerebral function whereas patients with encephalitis do. However, the distinction between the two entities may be blurred since some patients may have both a parenchymal and meningeal process with clinical features of both.

Viral encephalitis can be either primary or postinfectious. Primary infection is characterized by viral invasion of the CNS. Postinfectious encephalitis (also called acute disseminated encephalomyelitis) is thought to be an immune-mediated disease and here a virus cannot be detected or recovered and the neurons are spared.

A wide variety of different viruses can infect the CNS. Most viruses are capable of causing either meningitis or encephalitis, although in general, a given virus is more likely to cause one syndrome than the other. Some viruses are also associated with postinfectious encephalitis.

Some common causes of viral encephalitis include:

- HSV type 1.
- St. Louis encephalitis.
- West Nile virus.

Serologic testing is most important for patients who are not improving and who do not have a diagnosis based upon CSF analysis, culture, and PCR. Most viral etiologies require paired sera for diagnosis.

Brain biopsy can be considered in the patient if the etiology of encephalitis is still unknown.

CNS Tuberculosis

CNS TB may result in tuberculous meningitis, intracranial tuberculoma, and spinal tuberculous arachnoiditis.

Tuberculous meningitis accounts for about 1% of all cases of TB and 5% of all extrapulmonary disease in immunocompetent individuals. Early recognition of tuberculous meningitis is of importance as the clinical outcome depends greatly upon the stage at which therapy is initiated. Empiric antituberculous therapy should be started immediately in any patient with meningitis syndrome and CSF findings of low glucose concentration, elevated protein, and lymphocytic pleocytosis if there is evidence of TB elsewhere, either clinically or historically. Serial examination of the CSF by acid-fast stain and culture should be performed. Smears and cultures may yield positive results even days after treatment have been initiated. Nucleic acid amplification testing also may be helpful.

Tuberculomas are granulomatous foci within the brain parenchyma. They develop from coalescing tubercles acquired during an earlier period of hematogenous bacillemia. Centrally located lesions may reach considerable size without producing meningeal inflammation. Clinically silent single or multiple nodular enhancing lesions are commonly seen in the setting of meningitis; occasionally, they are seen in patients with miliary TB and no meningitis. These lesions generally disappear on therapy but may heal with calcification.

Spinal tuberculous arachnoiditis is seen most commonly in endemic areas. The pathogenesis is similar to that of meningitis, with focal inflammatory disease at single or multiple levels leading to gradual encasement of the spinal cord by a gelatinous or fibrous exudate.

KEY POINTS

- Culture of blood is considered to be the most definitive method for detection of bacteremia or fungemia. It is very important that the cultures are obtained before antimicrobial therapy is started.

- Prior to initiation of antimicrobial therapy, at least two sets of blood cultures taken from separate venipuncture sites should be obtained. One set consists of one aerobic and one anaerobic bottle.
- When a blood culture bottle turns positive, then a Gram stain is performed which is reported immediately.
- Blood culture bottles are routinely held for 5 days.
- Without appropriate treatment mortality of IE approaches 100%.
- Staphylococcal IE is a common cause of healthcare-related IE; streptococcal IE is a common cause of community-acquired IE.
- *S. aureus* is the most common bacterium infecting adult joints, followed by streptococci.
- Septic arthritis due to Gram-negative bacilli may be seen in the setting of trauma, intravenous drug users, neonates, older adults, and in association with underlying immunosuppression. In sickle cell patients infection with *Salmonella* may be seen. In neonates *Kingella* associated joint infections may be seen with a predilection for involvement of ankle joints.
- *P. acnes* is a relatively common cause of prosthetic joint infection following shoulder arthroplasty but is a rare cause of infection following hip or knee arthroplasty.
- Pneumonia due to *S. pneumoniae*: Predisposing causes include (1) chronic diseases, CHF, COPD, and diabetes, (2) immunodeficiency, and (3) absent spleen.
- Pneumonia due to *H. influenzae*: Predisposing causes include (1) COPD and (2) cystic fibrosis.
- Pneumonia due to *M. catarrhalis*: Predisposing cause include COPD.
- Pneumonia due to *S. aureus*: Predisposing causes include (1) viral respiratory illness (e.g., measles and influenza) and (2) IV drug abuse. It is an important causative agent for nosocomial pneumonia and has high incidence of complications.
- Pneumonia due to *K. pneumoniae*: Predisposing causes include (1) alcoholics and (2) debilitated and malnourished states.
- In atypical pneumonia classical features of pneumonia may not be seen. The patient lack features of consolidation as inflammation is mainly interstitial. The patient has more symptoms than signs. Leukocytosis may not be marked. Standard antibiotics used in community-acquired pneumonia will not be effective.
- In pulmonary TB, organism gains entry through inhalation (from another individual with active TB). Less than 5% of individuals with primary TB develop secondary TB.

- In HIV positive individuals, manifestation of TB depends largely on the CD4 cell count. If CD4+ counts >300 cells/mm^3, usual secondary TB ensues. If the CD4+ <200, then there is more likely development of progressive pulmonary TB (noncavitary consolidation). In such cases the sputum may be negative for AFB. Tuberculin skin test will be negative due to immune anergy and lesions may lack granulomas.
- Histoplasmosis, Coccidioidomycosis, and Blastomycosis are typically localized disease in immunocompetent individuals and disseminated disease in immunocompromised state. These are dimorphic fungi. They are in mold form in soil and in humans they are in yeast form. In culture we identify the mold form. In tissue or body fluids we identify the yeast form.
- A CD4 count of 200 or below (/mm^3) is a strong predictor of *P. jiroveci* pneumonia.
- More than 85% of cases of UTI are due to organisms that are normally found in the gastrointestinal tract.
- *E. coli* is the most common bacterial cause of UTI; it accounts for approximately 80% of UTI in children.
- Acute pyelonephritis is a serious complication of UTI and repeated attacks may lead to chronic pyelonephritis.
- Fulminant hepatic failure develops in less than 1% of patients with hepatitis A.
- HBsAg has several antigenic determinants. These are a, d, y, w, and r. Combinations of these determinants are used to classify HBV genotypes.
- HBsAg: Appears in blood in about 1–10 weeks after infection and then disappears; if HBsAg persists then this indicates a carrier state or chronic infection.
- HBeAg: Becomes present with acute infection and then disappears; if persists implies persistent viral replication and patients are infectious.
- HBcAg: Not tested.
- Anti-HBc: First antibody to appear; may be the only marker for hepatitis infection when HBsAg is negative and anti-HBs has not appeared.
- Anti-HBe: Second antibody to appear; implies decreased infectivity.
- Anti-HBs: Last antibody to appear; implies immunity.
- HBV DNA: Found in serum and liver; implies viral replication.
- Hepatitis C can cause acute hepatitis or chronic liver disease. Six different genotypes are identified. Type 1a and 1b account for 70% of cases in the United States.

- Most (90%) acute infections with hepatitis C are asymptomatic. About 90% of asymptomatic patients may develop chronic liver disease in the future. If the acute infection is symptomatic a higher percentage of these patients clear the virus. About 50%–75% of this group may develop chronic liver disease.
- Hepatitis D is an incomplete RNA particle enclosed in a shell of HBsAg. Hepatitis D infection can occur as co-infection with hepatitis B or as superinfection in an individual already infected with hepatitis B. Hepatitis D essentially makes outcome of hepatitis B worse.
- Hepatitis E is an RNA virus which causes hepatitis similar to hepatitis due to A. It is enterally transmitted and does not cause chronic liver disease.
- Invasive diarrhea, or dysentery, is defined as diarrhea with visible blood or mucus, in contrast to watery diarrhea.
- Most cases of acute diarrhea are mild or transient and do not require work up. However, individuals with bloody diarrhea, fever, and significant abdominal pain should be investigated.
- Causes of visible bloody diarrhea: EHEC, *Campylobacter*, *Salmonella*, and *Shigella*.
- *Campylobacter* infections are associated with reactive arthritis and Guillain Barre Syndrome.
- Salmonellae cause a broad range of infections, including gastroenteritis, enteric fever, and focal infections such as osteomyelitis.
- Salmonella may cause typhoid or enteric fever by infections with *S. typhi* or *S. paratyphi*. *Salmonella* can also cause gastroenteritis by the nontyphoidal Salmonella.
- Reactive arthritis and hemolytic-uremic syndrome (it is most often due to infection with enterohemorrhagic *E. coli* (particularly type O157:H7), but may also be due to *S. dysenteriae*).
- *C. difficile* is a spore-forming, toxin-producing, Gram-positive anaerobic bacterium that causes antibiotic-associated colitis. It colonizes the intestinal tract after the normal gut flora has been altered by antibiotic therapy. CDI is one of the most common healthcare-associated infections and a significant cause of morbidity and mortality among older adult hospitalized patients.
- The common causes of community-acquired bacterial meningitis in adults in are *S. pneumoniae* and *N. meningitides*.
- The usual causes of healthcare-associated bacterial meningitis are staphylococci and aerobic Gram-negative rods.

- *N. meningitidis* infections may cause petechiae and purpura. Arthritis is also another feature associated with such infections.
- Meningitis may be aseptic and this is when routine bacterial cultures are negative.
- Individuals with meningitis do not have abnormal cerebral function whereas patients with encephalitis do. However, the distinction between the two entities may be blurred since some patients may have both a parenchymal and meningeal process with clinical features of both.
- Viral encephalitis can be either primary or postinfectious.
- CNS TB may result in tuberculous meningitis, intracranial tuberculoma, and spinal tuberculous arachnoiditis.

REFERENCES

[1] Nielsen SL. The incidence and prognosis of patients with bacteremia. Dan Med 2015;62: pii B 5128.
[2] Ruhnke GW, Coca-Peraillon M, Kitch BT, Cullen DM. Marked reduction in 30 day mortality among elderly patients with community-acquired pneumonia. Am J Med 2011;124:171–8.
[3] Taylor LE, Foont JA, DeLong AK, Wurcel A, et al. The spectrum of undiagnosed hepatitis C virus infection in a US HIV clinic. AIDS Patient Care STDs 2014;28:4–9.
[4] Li KS, Sexton DJ, Mick N, Nettles R, et al. Proposed modification to the Duke criteria for diagnosis of infective endocarditis. Clin Infect Dis 2000;30:633–8.
[5] Watkins RR, Lemonovich TL. Diagnosis and management of community acquired pneumonia in adults. Am Fam Physicians 2011;83:1299–306.
[6] Fares A. Seasonality of hepatitis: a review update. J Family Prim Care 2015;4:96–100.

CHAPTER 4

Media for the Clinical Microbiology Laboratory

Contents

INTRODUCTION

Specimens sent to the microbiology laboratory are processed onto agar plates and broth or chemically defined medium, dependent upon the type of clinical material collected. This ensures that each medium provides targeted nutritional enrichment for growth amplification of the bacteria suspected of causing disease. Assessment through selective growth allows determination of biochemical characteristics that aid in organism identification, while at the same time inhibiting the growth of unwarranted organisms. These growth characteristics, along with macroscopic traits which include Gram-stain reactivity, acid-fastness, and structure recognition (presence of endospores, granules, and capsules) greatly assist in the identification process.

It is also imperative that detection of organisms in specimen samples take into account the necessity for oxygen. Specifically, bacteria respond to oxygen in various ways. Obligate aerobes require oxygen for growth and use O_2 as a final electron acceptor in aerobic respiration. These organisms are typically grown in conditions of 5%–10% CO_2 with the remainder oxygen [1]. When grown in a thioglycolate broth they will grow at the top where the highest concentration of oxygen exists [2]. Obligate anaerobes do not need or use oxygen as a nutrient; oxygen will act as a toxic substance and not permit growth of these organisms so they therefore grow at the bottom of a broth tube where oxygen is in the lowest concentration.

Microbiology and Molecular Diagnosis in Pathology.
DOI: http://dx.doi.org/10.1016/B978-0-12-805351-5.00004-1

These obligate anaerobic bacteria live by fermentation, anaerobic respiration, or photosynthesis and grow in an environment of mostly nitrogen gas with a small amount of hydrogen or carbon dioxide gas. Oxygen is removed using a palladium catalyst. In contrast facultative anaerobes can live by either aerobic or anaerobic types of metabolism. Under anaerobic conditions, they use anaerobic respiration, and in the presence of oxygen, they use aerobic respiration. These organisms grow near the top of the tube of broth. Finally, a group of aerotolerant anaerobes requires an exclusively anaerobic metabolism, yet they are indifferent to the presence of oxygen and can grow anywhere in the tube of broth media [3].

MEDIA USED IN MICROBIOLOGY LABORATORY

Most specimens are plated onto blood agar plates (BAP), Chocolate, Colistin-nalidixic agar (CNA), and MacConkey to allow for growth of most organisms and to select for Gram-positive organisms and Gram-negative organisms and to differentiate the Gram positives by hemolysis patterns and the Gram negatives by lactose fermentation. This is critical since further identification methods including biochemical testing and antimicrobial susceptibility testing require the use of a pure culture of an organism. Other selective media are used to grow fastidious organisms that would otherwise be overgrown by normal flora organisms that are in higher numbers and grow more rapidly. Media discussed in this chapter can be found in the Manual of Clinical Microbiology and the Difco & BBL Manual [4,5]. Common media used for growth of microorganisms are listed in Table 4.1.

Tryptic soy agar contains soybean meal and casein. It is typically combined with 5% sheep blood or Columbia agar with 5% defibrinated sheep blood. While it allows growth of most bacteria, the combination with sheep blood permits phenotypic differentiation of bacteria based on their hemolysis patterns. The patterns of hemolysis include three categories: Partial, full, or no hemolytic pattern. Alpha hemolysis which is only partial hemolysis of the blood and gives a dark greenish hue, full hemolysis or beta hemolysis which is full hemolysis of the agar and appears as a clearing of the blood, and gamma hemolysis which is no hemolysis. These plates are often referred to as BAP.

Chocolate agar is a variant of the blood agar described above, and is made using blood cells that have been lysed by elevated temperatures. Chocolate agar is a brown color and will grow all bacteria, including

Table 4.1 Common media used for growth of microorganisms

Media	Growth of organism	Nutritional/ selective/ differential	Aerobic/anaerobic
Blood agar	Gram positive, Gram negative	Nutritional	Aerobic/anaerobic
Chocolate agar	Gram positive, *Neisseria gonorrhoeae, Haemophilus influenzae*	Nutritional, XV factors	Aerobic/anaerobic
MacConkey agar	Gram negative	Selective/ differentiates lactose fermenters from nonlactose fermenters	Aerobic
Brain heart infusion agar	Gram positive, Gram negative	Nutritional	Aerobic
Eosin-methylene blue agar	Gram negative	Selective/ differentiates lactose fermenters from nonlactose fermenters	Aerobic
Phenylethyl alcohol agar	Gram positive	Selective	Aerobic
Colistin-nalidixic acid agar	Gram positive	Selective	Aerobic/anaerobic
Thayer–Martin/ Jembec agar	Gram positive, *N. gonorrhoeae, H. influenzae*	Nutritional, XV factors	Aerobic
PC agar	*Burkholderia cepacia*	Selective	Aerobic
Hektoen enteric agar	Gram negative	Selective/ differential for *Salmonella* and *Shigella*	Aerobic
CDC BAP	Obligate anaerobes	Selective	Anaerobic
K-V agar	Gram negative anaerobes	Selective	Anaerobic
LJ agar	Acid fast organisms	Selective	Aerobic
Middlebrook agar	Acid fast organisms	Selective	Aerobic
Egg yolk agar	Gram positive, Gram negative	Nutritional/ differential	Aerobic

(*Continued*)

Table 4.1 Common media used for growth of microorganisms (Continued)

Media	Growth of organism	Nutritional/selective/differential	Aerobic/anaerobic
Cefsulodin, Irgasan, Novobiocin agar	*Yersinia enterocolitica*	Selective	Aerobic
Campylobacter agar	*Campylobacter* species	Selective	Aerobic
Brain heart infusion broth	Gram positive, Gram negative	Nutritional	Aerobic/anaerobic
Tryptic soy broth	Gram positive, Gram negative	Nutritional	Aerobic/anaerobic
Thioglycolate broth	Gram positive, Gram negative	Nutritional	Aerobic/anaerobic
Chopped-meat broth	Gram positive, Gram negative	Nutritional	Anaerobic
LIM broth	Group B streptococci	Selective	Aerobic
Inhibitory mold agar	Mold	Selective	Aerobic
Birdseed agar	*Cryptococcus neoformans*	Selective/differential	Aerobic

Gram-positive and Gram-negative organisms including fastidious organisms that will not grow on BAP. The primary ingredient is beef extract which provides nitrogenous nutrients, vitamins, and minerals. *Neisseria gonorrhoeae* and *Haemophilus* spp. are two organisms that require the nutrients released by lysing the blood.

Brain heart infusion (BHI) is a nutritious medium used for the growth of a wide variety of organism types, including bacteria, yeasts, and filamentous fungi. The media contains casein and meat peptones that provide amino acids and peptides, peptones, and dextrose.

Thayer–Martin/Martin Lewis/Jembec agar contains hemoglobin which provides X factor, hemin, improving the growth of *Neisseria* species. This agar is also enriched with a supplement which provides V factor, nicotinamide adenine dinucleotide, which improves the growth of *Haemophilus*. Antibiotics inhibit normal genital flora so this media is used to grow *N. gonorrhoeae* from specimens likely contaminated with genital or rectal flora.

Phenylethyl alcohol with 5% defibrinated sheep blood or CNA Columbia agar with 5% defibrinated sheep blood is useful for cultivation of Gram-positive cocci from complex clinical specimens. These selective media inhibit most Gram-negative bacteria, and allow growth of *Staphylococcus* and *Streptococcus*.

Solid media such as anaerobic BAP, referred to as CDC BAP, is an enriched medium with additional vitamin K and hemin that is used for the growth of obligate anaerobes. These organisms include *Prevotella*, *Porphyromonas*, and the *Bacteroides fragilis* group. A media with gentamicin or kanamycin plus vancomycin (GV and KV) is used to select for these Gram-negative anaerobes and inhibit Gram-positive anaerobes.

Mueller–Hinton agar is a nonselective and nondifferential medium. This allows all organisms to grow and is the medium used for antibiotic susceptibility testing. This medium can be supplemented with blood or chocolate to allow for susceptibility testing of fastidious organisms. Since it is only used for susceptibility testing it is poured at the appropriate depth for that testing.

Selective and Differential Media

MacConkey agar not only selects for Gram-negative organisms by inhibiting Gram-positive organisms and yeast but also differentiates the Gram-negative organisms by lactose fermentation. Lactose ferments will stain pink while the nonlactose ferments will be clear colonies. The media also has the added advantage of inhibiting the swarming of *Proteus*.

Eosin-methylene blue is another media that electively grows Gram-negative organisms, while at the same time inhibiting Gram-positive organisms and some yeast by making use of toxic dyes. Enteric bacteria will vary in appearance on this media depending on their ability to ferment lactose and sucrose fermentation. *Escherichia coli* will have a green metallic sheen.

Additional media selective for Gram negatives but specifically designed to differentiate stool pathogens includes Hektoen enteric (HE) agar which contains bile salts. In addition to selecting for Gram-negative organisms the media also allows for the presumptive differentiation of *Salmonella* and *Shigella from* other Gram-negative organisms such as *E. coli*. *Escherichia coli* and other lactose ferments will produce yellow or orange colonies. Nonlactose fermenters including *Shigella* produce green colonies while *Salmonella* appears as black colonies due to production of hydrogen sulfide. Other organisms produce hydrogen sulfide and appear as black colonies on HE agar including *Citrobacter* spp.

Other Selective Agars for Specific Pathogens

Thiosulfate-citrate-bile salts-sucrose is a selective medium used to isolate *Vibrio cholera* and *Vibrio parahaemolyticus*. This agar contains a high concentration sodium thiosulfate and sodium citrate that inhibit the growth of Enterobacteriaceae. Gram-positive bacteria are inhibited by the addition of a pure bile salt. Sodium thiosulfate allows for the detection of hydrogen sulfide production and sucrose is used as a fermentable carbohydrate. *Vibrio cholera* will grow as large yellow colonies, while *V. parahaemolyticus* colonies will appear blue or green.

Cefsulodin, Irgasan, Novobiocin agar is a selective medium that isolates *Yersinia enterocolitica*. It is selective against the growth of other Gram-negative bacteria. *Aeromonas* species can also be isolated on this medium. *Yersinia enterocolitica* appears as small red target colonies.

Campylobacter BAP is a BAP that is supplemented with antibiotics and antifungal agents and is needed to isolate *Campylobacter* species due to their fastidious nature and the likelihood of overgrowth from normal flora in stool specimens.

Sorbitol MacConkey is a variant of MacConkey used for the selection of *E. coli* O157:H7. All *E. coli* ferment lactose and appear as pink colonies on MacConkey agar. However, the only serotypes of *E. coli* that can ferment sorbitol are *E. coli* O157:H7, the agent is most often implicated in hemolytic uremic syndrome and is therefore important to differentiate from the other normal flora *E. coli* in stool of patients with bloody diarrhea.

PC agar is a selective medium that allows for the growth of *Burkholderia cepacia*. It is often plated directly on respiratory specimens from patients with cystic fibrosis where this organism is a common pathogen. This media contains inorganic salts and proteinaceous substrates as well as crystal violet and bile salts, an inhibitor of Gram-positive cocci. Ticarcillin and polymyxin B are Gram-negative inhibitors. *Burkholderia cepacia* will have small pink colonies on this media.

Media containing charcoal is useful for growth of fastidious Gram-negative organisms. The charcoal is added to the media to bind toxins that might be in the specimen and that could inhibit their growth. Buffered charcoal yeast extract is one such medium which is particularly good for growth of *Legionella*. It utilizes yeast extract as a protein source, with no added starch, and can allow a more rapid growth of *Legionella pneumophila* for macroscopic evaluation of colonies. *Francisella tularensis* will also grow well on this media. Other charcoal containing media include Bordet

Gengou or Regan Lowe agars used to grow *Bordetella pertussis* species from the upper respiratory tract.

Media such as Egg yolk agar is used as a biochemical to help identify organisms more specifically. Egg yolk is an enriched medium that is differential for the detection of lecithinase and lipase production. Organisms that produce the enzyme lecithinase break down lecithin which results in a white opaque zone of precipitation that spreads beyond the edge of the colony. *Bacillus cereus and Bacillus anthracis* produce lecithinase, while *other Bacillus species considered normal flora do* not. *Clostridium perfringens* also produces lecithinase while the other *Clostridium* species do not. Fusobacterium produce lipase which gives sheen to the agar.

Chromogenic media contains substrates in the agar and takes advantage of the biochemical properties of bacteria. The organisms will produce the enzymes which will in turn cleave the chromogenic conjugate and release a chromophore. This allows for a visible color change in the colony and can be differentiated for isolation and differentiation. Various media have been developed to test bacteria and yeast. This media can be used to differentiate methicillin resistant *Staphylococcus aureus* from other bacteria, can differentiation *Candida* species, and can isolate Gram-negative bacteria that have a reduced susceptibility to carbapenem agents, among others.

Löwenstein–Jensen (LJ) medium, an egg-based media, is a particularly good medium for isolating acid fast organisms, like the *mycobacterium* species. On agar, it is easy to identify colony morphology as "smooth" or "rough," thus aiding in preliminary virulence identification. This agar contains the dye malachite green, a natural inhibitory dye towards general bacterial growth. Another agar used to culture *mycobacterium* species is Middlebrook agar which can be either selective or nonselective. Due to its clear color and the fact that it is more chemically defined than LJ medium it is used for susceptibility testing of *mycobacterium tuberculosis*. In order to grow mycobacterium more rapidly specimens are inoculated into a Middlebrook broth media and then put into an automated instrument for growth and detection. Although yeast such as *Candida* species grow on routine media such as BAP or Chocolate agar selective and nonselective media are also used to isolate fungal organisms from patient specimens.

Sabouraud dextrose agar is a selective medium primarily used for the isolation fungi and yeast but can also grow filamentous bacteria such as *Nocardia*. The low pH in this medium, pH 5.0, inhibits the growth of bacteria but allows the growth of yeast and most filamentous fungi.

BHI with 10% sheep blood contains defibrinated sheep blood and provides growth factors for the more fastidious fungal organisms. Two antibiotics are added to this media including chloramphenicol, a broad spectrum antibiotic that inhibits a wide range of Gram-positive and Gram-negative bacteria. As well as gentamicin, an aminoglycoside antibiotic that inhibits the growth of Gram-negative bacteria.

Inhibitory mold agar is a selective medium for the cultivation of pathogenic and nonpathogenic fungi. The media contains two peptones, derived from casein and animal tissue, and yeast extract. This media also contains dextrose; starch and dextrin are energy sources for the metabolism of fungi. Sodium chloride and metallic salts provide essential electrolytes and minerals. The antibacterial compound, chloramphenicol, is a broad spectrum antibiotic that inhibits a wide range of Gram-positive and Gram-negative bacteria. Other selective media include Mycosel media, which incorporates both chloramphenicol and cycloheximide. Cycloheximide, useful for growing dermatophytes, inhibits saprophytic yeast and molds. Nonselective media must also be included since cycloheximide can also inhibit pathogenic fungi such as *Cryptococcus*, *Aspergillus*, and *Zygomyces*.

Birdseed agar can be used for the identification of *Cryptococcus neoformans* due to the presence of caffeic acid, which is broken down by the production of phenol oxidase by the organism. Cryptococci appear as brown colonies on this clear agar.

Triple sugar iron (TSI) agar is a differential growth medium to identify organisms based on properties related to fermentation of specific carbohydrates, and ability to reduce sulfur. TSI agar contains 0.1% glucose, 1% lactose, 1% sucrose, and 2% peptone. The fermentation of glucose will produce acid, changing red agar to yellow due to reduced pH with the pH indicator, phenol red. TSI agar for identification of representative microorganisms are summarized in Table 4.2.

Glucose is at a low supply of 0.1% and after the supply is exhausted, the bacteria will catabolize the peptones and amino acids for energy. Alkaline end products produced convert the pH to alkaline and change the agar back to red. For example, *Salmonella* and *Shigella* attack glucose but not lactose or sucrose, producing an alkaline slant (red), and acid butt (yellow). Metabolism of glucose into carbon dioxide measures gas formation by visualization of bubbles or cracks in the agar. Sodium thiosulfate serves as the electron receptor for reduction of sulfur and produces hydrogen sulfide, which looks black on the agar slants [6].

Table 4.2 Triple sugar iron agar for identification of representative microorganisms

Organism	Triple sugar (slant/butt)	H_2S	Gas
Escherichia coli	A/A	−	+
Shigella dysenteriae	Alk/A	−	−
Salmonella	Alk/A	+	+
Klebsiella pneumoniae	A/A	−	+
Enterobacter cloacae	A/A	−	+
Serratia marcescens	Alk/A	−	+
Pseudomonas	Alk/Alk	−	−

A, acid, yellow; *Alk*, alkaline, red; *H_2S*, hydrogen sulfide; *Gas*, bubbles/cracks in agar.

BROTHS

Liquid media such as Schaedler broth is used on many wound and tissue specimens to allow for growth of fastidious organisms including facultative anaerobes. The broth also aides in diluting out any inhibitors in the specimen such as antimicrobials or white blood cells (pus) to allow for growth of organisms that may be in small numbers in the specimen.

BHI is a nutritional broth used for the cultivation of aerobic and anaerobic bacteria, yeast, and mold.

Tryptic Soy Broth is a medium that provides enzymatic digests of casein and soybean meal to provide amino acids and nitrogenous substances. This broth is used for the cultivation of aerobic and facultatively anaerobic bacteria.

Thioglycolate broth is an enriched differential medium used for the cultivation of bacteria, especially anaerobes and microaerophiles.

Chopped-meat broth is an enrichment broth medium that supports the growth of most anaerobes.

LIM broth is a selective medium for group B streptococci. CNA are added to inhibit Gram-negative organisms.

KEY POINTS

- Collected specimens are processed onto defined medium targeted to nutritional enrichment and amplification of the pathogens suspected of causing disease, over that of normal flora.
- Defined media may take the form of solid agar, liquid broth, or culture media.

- Selective growth allows determination of biochemical and growth characteristics. This, in combination with identifying macroscopic traits, assists in the identification process.
- Factors to consider in growth media include requirements for nutrients and oxygen, lytic capability of blood components, and factors that allow identification of fermentation processes.
- Specific stains allow differentiation of those organisms with or without certain cell wall constituents.
- Use of toxic dyes permits further identification of patterns specific to classes of organisms.
- Further assessment can be made using selective agars that demonstrate dependence on salts and carbohydrates (sugars).
- Enzyme production and biochemical assessment will also allow specificity of identification.

REFERENCES

[1] AGarcia LS, Isenberg HD, editors. Para technical processing of specimens for aerobic bacteriology. In: Clinical microbiology procedures handbook. 3rd ed. Washington, DC: ASM Press; 2010 [2007 update].
[2] Claros MC, Citron DM, Goldstein EJ. Survival of anaerobic bacteria in various thioglycolate and chopped meat broth formulations. J Clin Microbiol 1995;33:2505–7.
[3] Garcia LS, Isenberg HD, editors. Anaerobic bacteriology. In: Clinical microbiology procedures handbook. 3rd ed. Washington, DC: ASM Press; 2010 [2007 update].
[4] Jorgensen J, Pfaller MA, editors. Reagents, stains, and media: bacteriology. In: Manual of clinical microbiology. 11th ed., vol. 1. ASM Press; 2015. p. 316–49, Washington, D.C.
[5] Zimbro MJ, Power DA, Miller SM, Wilson GE, Johnson JA, editors. Difco™ & BBL™ manual. Manual of microbiological culture media. 2nd ed. Maryland: BD Diagnostics – Diagnostic Systems; 2009.
[6] Procop G, Koneman EW. Koneman's color atlas and textbook of diagnostic microbiology. Philadelphia: Lippincott Williams & Wilkins; 2016.

CHAPTER 5

Biochemical Tests and Staining Techniques for Microbial Identification

Contents

INTRODUCTION

Biochemical tests as well as the various stains used for microbial identification will be discussed in this chapter. Specific examples used for differentiation of microorganisms will be reviewed [1–4].

BACTERIAL STAINS

Named after Hans Christian Gram, the Gram stain is the first step for preliminary identification of bacteria and is used as a guide for empiric treatment. The Gram stain differentiates Gram-positive from Gram-negative bacteria and cocci from bacilli. The size of the organisms as well as the

Microbiology and Molecular Diagnosis in Pathology.
DOI: http://dx.doi.org/10.1016/B978-0-12-805351-5.00005-3

formation, pairs, chains, and clusters, can be very helpful in preliminarily identifying the organism. Gram-positive cells will stain purple due to the retention of the crystal violet from the thick peptidoglycan in their cell wall. Gram-negative cells will lose the crystal violet stain after the decolorizing agent is used and retain the safranin counterstain due to the thinner layer of peptidoglycan these cells have (Tables 5.1 and 5.2).

Gram Stain

Prepare smear from specimen or place a few drops from body fluid onto clean glass slide and fix on heat block at 65°C or until completely dry. Methanol fixation is preferred over heat fixation when staining slides prepared from positive blood culture bottles. Methanol fixation prevents lysis of blood cells resulting in a cleaner background. The steps include:

Step 1. Apply crystal violet to a heat fixed slide for 20 seconds.

Step 2. Add iodine, the mordant, to fix crystal violet for 20 seconds.

Step 3. Apply a decolorizing agent such as alcohol/acetone to remove the primary stain. Timing is dependent on the thickness of the specimen.

Step 4. Apply safranin, the counterstain, for 20 seconds.

Step 5. Rinse and allow to air dry. Observe with light microscope.

Fastidious bugs, such as *Brucella*, stain weakly, and the counterstain should be applied for a longer period of time, between 2 and 3 minutes. Acridine orange stain can also be used, for these fastidious Gram-negative organisms, which retain counterstain poorly. Acridine orange stain intercalates into the nucleic acid of the organism and stains the organisms orange.

Acid Fast Strain

Acid fast stains are used to differentiate acid fast organisms such mycobacteria. Acid fast bacteria have a high content of mycolic acids in their cell walls. Acid fast bacteria will be red, while nonacid fast bacteria will stain blue/green with the counterstain with the Kinyoun stain. The steps include:

Step 1. Apply carbol fuchsin to a fixed slide for 1 minute followed by rinsing.

Step 2. The decolorizing agent, 3% hydrogen chloride (HCl), is applied for 2 minutes and remove the primary stain and rinse.

Step 3. Apply the counterstain, methylene blue, for 2 minutes then rinse.

Step 4. Allow to dry and observe slide with a light microscope.

Ziehl Nielson can also be used to stain mycobacteria but uses heat while the Kinyoun method does not. The Kinyoun method can be modified as a weak acid fast stain, which uses 5% sulfuric acid instead of hydrochloric acid. The weak acid fast stain in addition to staining mycobacteria will stain organisms that are not able to maintain the carbol fuchsin after decolorizing with HCl, such as *Nocardia* spp., *Rhodococcus* spp., *Tsukamurella* spp., and *Gordonia* spp. The weak acid fast stain also helps differentiate among the organisms that appear as Gram-positive branching filamentous rods such as *Nocardia* spp. and *Streptomyces* spp. Nocardia will stain positive with a weak acid fast stain and *Streptomyces* spp. will not.

Rhodamine Auramine Strain

Rhodamine auramine stain is used for the detection of mycobacteria directly from clinical specimens. The dye binds with the mycolic acids and fluoresces under ultraviolet light. Acid fast organisms (mycobacteria) will appear yellow or orange under ultraviolet light. The fluorescent stain is more sensitive for the detection of mycobacteria but is not specific so the appearance of beaded rods is critical in considering the stain positive. The steps include:

Step 1. Apply rhodamine auramine to fixed slide for 5 minutes and then rinse.

Step 2. Apply the decolorizing agent, acid-alcohol, and then rinse.

Step 3. Apply potassium permanganate for 2 minutes and then rinse.

Step 4. Allow to dry and observe slide with a light microscope.

Flagellar Strain

The flagellar stain stains the bacterial flagella since they cannot be seen with a light microscope. Crystal violet is used in an alcoholic solution and after staining, the alcohol evaporates, enhancing the presence of the flagella. Flagella will stain purple and should be examined with oil immersion. The steps include:

Step 1. Prepare slide by placing a colony of the bacterial isolate in a drop of deionized water and let it air dry.

Step 2. Flood the slide with the flagella stain and then rinse.

Step 3. Air dry and examine with light microscope.

Endospore Stain

The endospore stain is used to differentiate between vegetative bacteria and cells that have produced endospores. These spores are produced by the bacteria to protect themselves from lack of nutrients and unfavorable environmental conditions such as desiccation and extreme temperatures. Endospores will appear green and vegetative cells will stain pink. The steps include:

Step 1. Apply malachite green for 30 minutes and then rinse.

Step 2. Apply safranin for 1–5 minutes and then rinse.

Step 3. Allow to dry and observe slide with a light microscope.

FUNGAL STAINS

Lactophenol cotton blue is a stain that is used to examine fungal elements following either a tape preparation or a scraping. This stain contains phenol, which will kill the organisms, lactic acid which preserves fungal structures, and cotton blue which stains the chitin found in the fungal cell walls. The microscopic fungal morphology will be used to identify the mold. The steps include:

Step 1. Apply lactophenol blue onto a clean slide.

Step 2. Using clear tape adhesive, light touch the fungal hyphae on the plate so that the fungal elements stick to the tape. Alternatively fungal growth can be scraped from an agar medium and mixed with the lactophenol blue dye and the slide can be cover-slipped and viewed under the microscope.

Step 3. Place the tape onto the slide with lactophenol blue and observe under a light microscope.

Microscopic morphology of the filamentous fungus is critical in the identification of molds since biochemical and molecular methods are not readily available for the identification of molds. Hyaline fungi will pick up the blue color of the lactophenol, while the dematiaceous fungi will stain brown due to the pigment produced by the mold.

India Ink

India ink is performed on cerebral spinal fluid to diagnose cryptococcal meningitis. The *Cryptococcus* yeast cell wall is composed of a polysaccharide capsule. India ink will stain only the background and the extracellular capsule will not stain and appears as a halo surrounding the yeast. A cryptococcal antigen test is available in serum and cerebral spinal fluid (CSF). This test is more sensitive and quantitative, but is usually not offered as a stat test as the India ink.

Caution must be taken in reading the slides since WBCs can appear similar in size and shape as *Cryptococcus*. Make sure to confirm the presence of budding yeast. The steps include:

Step 1. Prepare wet mount with CSF.

Step 2. Apply India ink.

Step 3. Mix gently so that the dye is not too thick and place coverslip.

Step 4. Observe slide with a light microscope.

Potassium hydroxide (KOH) preparation can be used to differentiate dermatophytes and *Candida albicans* from skin disorders such as eczema.

KOH is an alkali that will digest skin, hair, and nails allowing fungal elements to be seen under a light microscope. The steps include:

Step 1. Prepare a wet mount of the specimen and apply KOH.

Step 2. Place a coverslip over the specimen.

Step 3. Observe slide with a light microscope.

Calcofluor white stain is a fluorescent stain that binds to cellulose and chitin which is found in the cell walls of yeast and molds. Fungal and parasitic organisms, specifically *Pneumocystis jirovecii* will appear fluorescent green or blue under ultraviolet light. This is the stain used directly on clinical specimens. This stain is the most sensitive for visualizing few fungal elements. Slides should be maintained in the dark. The steps include:

Step 1. Prepare a wet mount of the specimen and apply Calcofluor white stain and 10% KOH.

Step 2. Place a coverslip over specimen.

Step 3. Observe slide after 1 minute under ultraviolet light.

Direct Fluorescent Antibody of *Pneumocystis jirovecii*

The direct fluorescent antibody assay for *P. jirovecii* is prepared on bronchial alveolar lavages. Specimens are treated with enzymes to remove host debris and fixed onto a slide. An antibody to *P. jirovecii* and a fluorescently labeled antimouse antibody are added and incubated. A fluorescence microscope is used to identify the presence of oocysts that appear bright green in clusters or in a honey comb pattern.

PARASITE STAINS

Wheatley's modification of the Gomori trichrome stain, or trichrome stain for short, is a rapid staining procedure that provides visualization of the parasites allowing differentiation by morphology and internal structure. The preparation requires incubation in the trichrome stain for 10 minutes among other washes. Intestinal protozoan cysts and trophozoites are visible as pink in liquid stool specimens, while the background will appear blue due to the counterstain, Ryan Stain.

Performed from PhotoFix as direct stain or on concentrated material.

Modified Trichrome Stain

Modified trichrome stain is used to identify intestinal microsporidia. Stool specimens can be prepared with trichrome stain as above, though it is 10 times the amount used and the staining time is increased to 90 minutes.

Microsporidia are visible as pink structures in liquid stool specimens, while the background will appear blue due to the counterstain, Ryan Stain. Due to the small size, 1–2 μm, electron microscopy from gastrointestinal (GI) biopsy is recommended for diagnosis of microsporidium.

A direct stain can be prepared on fresh stool sent to the lab looking for eggs or microfilariae. A drop of iodine is added to a wet mount, killing the organism and staining the internal structures. A coverslip is placed onto the slide and visualized with a low power objective. Organisms identified include *Giardia lamblia*, *Ascaris*, pin worm, and hookworm in addition to others.

Modified Acid Fast Strain

A modified acid fast stain can be used to stain oocysts of the coccidia: *Cryptosporidium*, *Cystoisospora*, and *Cyclospora*. This stain is similar to the stain used for mycobacteria (above), but can also use 3% malachite green dye as the counterstain. The oocysts will appear red as they retain the red dye from the carbol fuchsin. The steps include:

Step 1. Apply carbol fuchsin to a fixed slide for 1 minute and then rinse.

Step 2. The decolorizing agent, 3% hydrogen chloride (HCl), is applied for 2 minutes and remove the primary stain and then rinse.

Step 3. Apply the counterstain, methylene blue, for 2 minutes and then rinse.

VIRAL STAINS

Although diagnosis of most viral infections is made by molecular detection techniques viruses can also be grown in cell culture and identified by their cytopathic effect (CPE). Preliminary identification of the virus can be made by the presence of nuclear or cytoplasmic inclusions and syncytia following incubation on tissue culture cells either in tubes or more commonly in shell vials. Specific identification can be made by the addition of an antibody-based stain to the coverslip in the shell vial following several days of incubation.

Antibody-Based Immunostains

The specimen is inoculated onto the cells specific for the viruses and then spun at a low speed. Examples of stains include peroxidase- or fluorescein isothiocyanate (FITC)-labeled antibodies that use substrates such as a β-galactosidase. The substrate, β-galactosidase, is detected and will appear

Table 5.1 Common stains used for organism differentiation

Bacterial stains	Examples of organisms stained	
	Positive	**Negative**
Gram stain	*Staphylococcus aureus*	*Escherichia coli*
Acid fast stain	*Mycobacteria tuberculosis*	*Bacillus cereus*
Modified acid fast stain	*Nocardia* spp.	*Streptomyces* spp.
Rhodamine auramine	*Mycobacteria tuberculosis*	*S. aureus*

Fungal stains	Description of organisms
Lactophenol blue	Hyphal elements stain blue, dematiaceous mold appear brown
India ink	Cryptococci have a capsule and will appear as a halo surrounding the yeast
Potassium hydroxide (KOH)	KOH will degrade keratinized cells, and fungal elements are apparent in hair, skin, and nails
Calcofluor white	Fungal elements appear fluorescent under UV light
Direct fluorescent antibody of *Pneumocystis jirovecii*	*Pneumocystis* oocysts appear fluorescent green with a specific antibody-based stain
Parasite stains	
Trichrome stain	Protozoan trophozoites and cysts are seen
Modified trichrome stain	Microsporidia spores will stain pink/red
Iodine stain	Cyst glycogen stains reddish/brown, cytoplasm stains yellow, chromatin stains brown/black
Modified acid fast stain	Coccidia will stain pinkish-red
Viral stains	
Antibody based	Staining appears bluish when virus is present
Immunofluorescence	A fluorescent signal is visible in the presence of virus

blue after genetically activated and overproduced. A diagnosis of cytomegaly virus infection can be made relatively rapidly without waiting 1–3 weeks for the CPE to appear.

Biochemical Tests

Biochemical tests are often used to differentiate various microorganisms. These tests are based on the production of enzymes or lack thereof to visualize a biochemical change with a substrate and the colony of bacteria. Biochemical tests are relatively rapid and a small amount of bacteria from a colony growing on a plate can be used to obtain a result.

Table 5.2 Common biochemical tests available for the differentiation of bacteria [1–4]

Biochemical tests	Basis of test	Procedure	ID of organisms
Pyrolidonyl arylamidase (PYR)	The substrate is L-naphthylamide-β-naphthylamide which is hydrolyzed by PYR	Chromogenic solution mixed with a colony on a moistened filter paper. Incubate 2 min, positive will turn pink color	Positive result *Enterococcus* spp., *Staphylococcus lugdunensis*
Optochin	Optochin (ethylhydrocupreine hydrochloride)	Disk is impregnated with a solution of ethylhydrocupreine hydrochloride placed on a lawn of bacteria	*Streptococcus pneumoniae* is sensitive and a zone of inhibition forms around an optochin disk, *Viridans streptococci* are resistant and no zone is apparent
Bacitracin	Bacitracin is a bactericidal drug	Disk is impregnated with a solution of bacitracin placed on a lawn of bacteria	*Streptococcus pyogenes* is sensitive and a zone of inhibition forms around an bacitracin disk, other beta hemolytic streptococci are resistant and no zone is seen
Novobiocin	Novobiocin is a bactericidal drug	Disk is impregnated with a solution of novobiocin placed on a lawn of bacteria	*Staphylococcus saprophyticus* is resistant no zone of inhibition is seen around the disk. *Staphylococcus epidermidis* is susceptible and a zone of inhibition is seen
Trimethoprim/sulfamethoxazole (TMP/SXT)	TMP/SXT is a bactericidal combination drug	Disk is impregnated with a solution of TMP/SXT placed on a lawn of bacteria	Group A and Group B streptococci are resistant
Furazolidone (microdase)	Furazolidone is a nitrofuran antibiotic	Disk is impregnated with a solution of furazolidone placed on a lawn of bacteria	Staphylococci are susceptible while Micrococci are resistant

Urease	An enzyme that catalyzes the hydrolysis of urea into carbon dioxide and ammonia	Nutrients coupled with the use of pH buffers prevent all but rapid urease-positive organisms from producing enough ammonia to turn the phenol red pink	*Cryptococcus* spp., Brucella *Proteus mirabilis* and *Proteus vulgaris* are positive for urease
Indole	Indole is an organic compound produced by bacteria as a degradation product of the amino acid tryptophan	Tryptophan is hydrolyzed by tryptophanase end products, like indole. Indole production is detected by Kovac's or Ehrlich's reagent, and produces a red compound	*Escherichia coli* is positive for indole
Oxidase	Cytochrome oxidase enzyme catalyzes the oxidation of cytochrome c.	1% solution of tetramethyl-*p*-phenylene-diamine dihydrochloride is added to a filter paper and a colony of bacteria is smeared on filter paper. Positive oxidase production will turn filter paper purple	*Neisseria*, *Moraxella*, *Campylobacter*, *Pasteurella* species and *Pseudomonas* are oxidase positive
Catalase	Catalase is the enzyme that breaks hydrogen peroxide (H_2O_2) into H_2O and O_2	A drop of H_2O_2 is mixed with a colony of bacteria. If the bacteria produces catalase bubbles will form	*Staphylococcus species* are positive for catalase, *Streptococcus* spp. are negative
Coagulase	Coagulase is an enzyme that converts fibrinogen to fibrin, which means that it can coagulate plasma	Plasma is mixed with a colony of bacteria and read after 4 h. Coagulation of the plasma produces stable coagulate	*S. aureus* is coagulase positive while all other *Staphylococcus* spp. are negative

(*Continued*)

Table 5.2 Common biochemical tests available for the differentiation of bacteria [1–4] (Continued)

Biochemical tests	Basis of test	Procedure	ID of organisms
Latex agglutination	Latex agglutination determines the absence or presence of an antigen or antibody	A sample of specimen is mixed with latex beads coated with a specific antibody or antigen. If substance is present, latex beads will agglutinate	One test will differentiate *S. aureus* (positive) from other *Staphylococcus* spp. (negative). Can be used to differentiate *Streptococcus* spp.
Bile solubility	Deoxycholate can solubilize some bacteria	A drop of bile is dropped directly onto the colonies growing on BAP and are dissolved	*Streptococcus pneumoniae* are soluble in bile, *Viridans streptococci* are not soluble
Nitrate reduction	Organisms may reduce nitrate to nitrites	Nitrites are detected by adding α-naphthylamine and sulfanilic acid. A red color will appear in the presence of nitrites	Enterobacteriaceae demonstrate nitrate reduction
Bile esculin	Hydrolysis of esculin can occur in the presence of bile. The hydrolysis of esculin will react with ferric citrate and produce iron salts, blackening the medium	Bile salts are selective, while esculin is differential. The bacteria can be grown on this media	*Enterococcus* spp. demonstrate esculin hydrolysis and form blackening of the medium, while *Streptococcus* spp. are negative
6.5% NaCl	Some bacteria may survive in high concentrations of NaCl	Bacteria are grown in the presence of 6.5% NaCl in broth or media	*Enterococcus* spp. can grow in the presence of 6.5% NaCl and are positive for the bile esculin test

Methyl red	Bacteria can produce a red color in the presence of the pH indicator, methyl red. Acid is produced from dextrose fermentation	Bacteria are grown in a broth with methyl red. By-products from bacterial growth will decrease the pH of the broth and produce a red color	*Klebsiella* is negative and *E. coli* is positive
Voges–Proskauer (VP)	Bacteria produce neutral-reacting end products. Glucose fermenters further break down the chemicals in the VP broth and form a red complex	Bacteria are grown in a broth with VP broth. The change of broth to a red color is a positive test	*Klebsiella* and *Enterobacter* spp. are positive and appear red, *E. coli* is negative
Hugh–Leifson oxidative/fermentative (OF) medium	Broth will distinguish between anaerobic and aerobic breakdown of glucose. The media contains high concentration of carbohydrates and low concentration of peptic digest	Bacteria are grown in OF broth. Aerobic organisms produce acid in an unsealed tube with no growth and no acid in a sealed tube. Anaerobic organisms will produce acid in both sealed and unsealed tubes	*E. aerogenes*, *E. coli*, *Salmonella typhi* produce acid and gas production in sealed and unsealed tubes. *P. aeruginosa* produce acid only in an unsealed tube

(*Continued*)

Table 5.2 Common biochemical tests available for the differentiation of bacteria [1–4] (Continued)

Biochemical tests	Basis of test	Procedure	ID of organisms
Potassium hydroxide (KOH) 3% string test	A 3% solution of KOH will lyse Gram-negative cell walls. When the DNA is liberated, the mixtures turn "stringy"	A colony is mixed with a 3% KOH solutions	Gram negatives will result in a stringy mixture and are positive, Gram positives are negative for this test
Staphylococcus streak test	Some bacteria require X and V factor to grow. *S. aureus* will break down red blood cells in BAP releasing the X and V factors	A streak of *S. aureus* is made on a BAP, bacteria requiring X and V factor streaked near the *S. aureus* will grow in larger colonies	*Haemophilus influenzae* grows well near the *S. aureus* colonies

Biochemical tests are often used to differentiate various microorganisms. These tests are based on the production of enzymes or lack thereof to visualize a biochemical change with a substrate and the colony of bacteria. Biochemical tests are relatively rapid and a small amount of bacteria from a colony growing on a plate can be used to obtain a result.

KEY POINTS

- Stains, such as the Gram stain, are used to differentiate bacteria based on the composition of the cell wall.
- Special stains used to detect organisms directly from the specimen include the mycobacterial stain auramine rhodamine and calcofluor white for fungal organisms.
- After the bacteria grows as a colony on a culture plate, special stains such as the Kinyoun stain can be used to identify mycobacterium which are positive and differentiate them from *Nocardia* which are negative.
- Weak acid fast stain which uses sulfuric acid instead of hydrochloric acid as a decolorizer differentiates *Nocardia* which are positive from other Gram-positive bacilli.
- Mycobacterium species will also be positive with the weak acid fast stain.
- Weak acid fast can also be used to identify parasites such as *Cryptosporidium* and *Cyclospora*.
- Trichrome stains are useful to identify parasites directly in fecal specimens.
- Biochemical tests are generally fast and efficient way to differentiate bacteria. Examples of these biochemical tests include sensitivity to antibiotics, production of enzymes, and survival under various chemical conditions.
- Viruses can grow in cell culture and CPEs of the virus used to identify the virus, but molecular methods are now commonly used to diagnose viral infections due to the increased sensitivity and specificity as well as the rapidity of the tests.

REFERENCES

[1] Murray PR, Rosenthal KS, Pfaller MS. Medical microbiology. 8th ed. Philadelphia, PA: Elsevier; 2016.
[2] CLSI. Performance standards for antimicrobial susceptibility testing. CLSI supplement M100S, 26th ed. Wayne, PA: Clinical and Laboratory Standards Institute; 2016.
[3] Jorgensen JH, Pfaller MA, editors. Manual of clinical microbiology. 11th ed. Washington, DC: ASM Press; 2015.
[4] Garcia LS, editor. Clinical microbiology procedures handbook. 3rd ed. Washington, DC: ASM Press; 2010.

CHAPTER 6

Overview of Bacteria

Contents

Microbiology and Molecular Diagnosis in Pathology.
DOI: http://dx.doi.org/10.1016/B978-0-12-805351-5.00006-5

INTRODUCTION

Bacteria are among the most common organisms to cause infection in humans, both immunocompetent and those who are predisposed. While it is usually beneficial for bacteria to colonize humans as normal microbiota, they can occasionally find a route to foreign areas and cause disease [1]. Historically, bacteria have been classified according to staining methods, biochemical properties, and most recently, by molecular methods, especially 16S ribosomal RNA [2]. Bacteria vary in phenotype and genotype which can consequently be useful for differentiation.

This chapter discusses the following points which are necessary when diagnosing an infection and providing adequate antibiotic treatment:

- Basic physical, chemical, and structural characteristics.
- Epidemiology of each infection.
- Pathophysiology changes associated with infection, including host-related factors.
- Clinical manifestations of infection.
- The biochemical assays useful in differentiating genera from other closely related genera, and among the species.
- Most commonly used antimicrobial treatments.

DIAGNOSTIC APPROACH

Bacterial disease will vary by type of organism and the immune status of the host. Pathogenic bacteria have a wide range of virulence factors and will result in infections that vary in signs and symptoms. Clinical diagnosis will depend on multiple aspects of a patient's history, a physical examination, and results from patient's specimens sent to the microbiology laboratory for processing. Specimens collected should be representative of the infection and caution must be used to avoid contamination with normal flora. The microbiology laboratory will then process the specimens accordingly. Initial examinations include Gram stains for bacteria, followed by inoculation in aerobic and anaerobic media to allow for growth of the organisms.

GRAM-POSITIVE BACTERIA

Staphylococcus species are Gram-positive cocci arranged in clusters and are facultative anaerobes. These bacteria are comprised of a thin polysaccharide capsule which can protect them from phagocytosis in addition to a slime layer which is composed of monosaccharides, proteins, and small peptides that are useful for cell attachment and biofilm formation. These bacteria represent common skin colonizers which are opportunistic depending on the strain and route of infection.

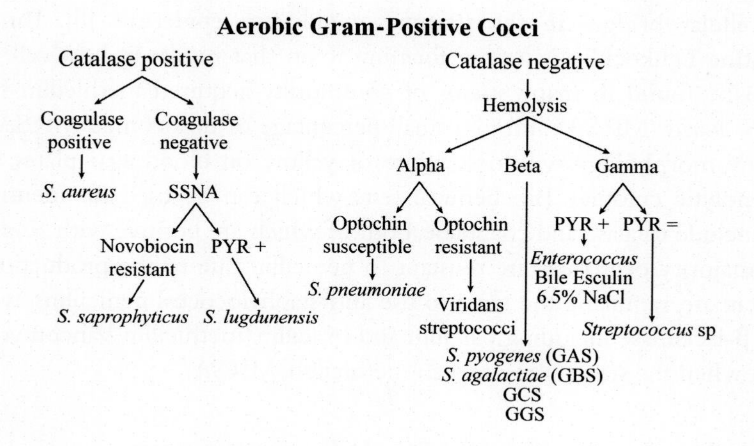

Coagulase Positive *Staphylococcus*

Staphylococcus aureus produce a carotenoid called staphyloxanthin, which results in yellow or golden colonies ("aureus" for golden) [3]. Frequent recurrence is seen likely due to the virulence factors and ability to evade innate and adaptive immune responses [4]. *Staphylococcus* species can cause cutaneous infections including boils, folliculitis, and impetigo. Systemic infections such as bacteremia, endocarditis, septic arthritis, osteomyelitis, toxic shock syndrome (TSS), scalded skin syndrome, and purpura fulminans are possible. Other diseases include food poisoning due to heat stable enterotoxins. These enterotoxins can be stable at temperatures over 100°C for up to 30 minutes, are resistant to acids, and require a very small amount of enterotoxin to cause intoxication [5]. There is a rapid onset of symptoms, about 2–6 hours after ingesting the food and has a quick resolution, within about 24 hours after onset. Symptoms include vomiting and large volumes of watery diarrhea (lacking blood, pus, fever, or severe abdominal pain). Transmission of staphylococci occurs through fomites, consumption of contaminated food, and direct contact with people who are colonized with this organism [6].

Toxins are highly involved in virulence of the bacteria. Cytotoxins, exotoxins, and enterotoxins including α-toxin/hemolysin can lyse many types of cells which destroy tissues. Toxic shock syndrome toxin-1 (TSST-1) is a super-antigen which can stimulate the release of cytokines. The toxin can penetrate mucosal barriers and spread systemically while the infection remains localized [7]. Exfoliative toxins are proteases that degrade intercellular bridges in the stratum granulosum epidermis [8]. Panton-Valentine Leukocidin is a pore forming toxin that results in host cell lysis. It can be found in many strains of community-acquired methicillin resistant *S. aureus* (MRSA) and in a small percentage of nosocomial MRSA [9]. Colony morphology will look white or yellow on blood agar plates with β-hemolytic colonies. Biochemical tests which can be used for identification include catalase and coagulase, both of which are positive with *S. aureus*. The majority of isolates are resistant to penicillin due to the production of β-lactamase, requiring the need to use anti-staphylococcal penicillins, which resist β-lactamase including nafcillin and oxacillin/methicillin. Vancomycin is given when the strain is resistant to methicillin, MRSA.

Staphylococcus species Not *Staphylococcus aureus*

Staphylococcus species not *S. aureus* (SSNA) strains are normally found on the skin and mucosal surfaces and can be associated with nosocomial infections, particularly in close proximity to artificial devices [10]. *Staphylococcus epidermidis* is the most common staphylococcal species and rarely causes infection. Hospital-related infections are typically line or device related due to proteins produced by the bacteria that allow for attachment to a foreign body in the patient which aids in the production of biofilms [11]. Two species of *Staphylococcus* which must be differentiated from the rest are *Staphylococcus saprophyticus* and *Staphylococcus lugdunensis*. The colony morphology on blood agar plates are white, nonhemolytic colonies. Biochemical tests that can be used to differentiate SSNA from *S. aureus* include a negative coagulase result. Infection can be treated with vancomycin, trimethoprim/sulfamethoxazole plus rifampicin, minocycline, linezolid, and daptomycin.

Staphylococcus saprophyticus

S. saprophyticus causes urinary tract infections in young women. Symptoms include dysuria, pyuria, and numerous organisms in the urine [12]. Biochemically they are resistant to the antibiotic novobiocin, which is distinguishable from other SSNA.

Staphylococcus lugdunensis

Staphylococcus haemolyticus and *S. lugdunensis* are associated with endocarditis, catheter and shunt infections, prosthetic joint infections, and arthritis [13]. Differentiation from other SSNA can be performed by the presence of pyrrolidonyl arylamidase (PYR). Catalase positive cocci are summarized in Table 6.1.

Streptococci

The streptococci are biochemically different from the staphylococci in regards to their lack of catalase activity. These bacteria are facultative anaerobes and appear as Gram-positive cocci in pairs and/or chains on a Gram stain. These bacteria can be characterized by the varying hemolytic processes such as alpha, greenish discoloration indication of partial hemolysis, beta, indicating complete hemolysis, and gamma, no hemolysis. Streptococci can also be classified by the genus and species of the organism or the serological analysis called Lancefield grouping. Streptococci are all sensitive to penicillin. If the patient has a penicillin allergy, clindamycin, vancomycin, and levofloxacin can be tested for sensitivities and used for treatment.

The large number of species that fit into this category has recently been rearranged into seven groups [14]. The Pyogenic cocci group

Table 6.1 Catalase positive cocci

Organism	Gram stain morphology	Colony on BAP	Hemolysis	Coagulase	Pyrrolidonyl arylamidase	Novobiocin
Staphylococcus aureus	Cocci in clusters	Flat, golden yellow pigment	Beta	+	−	S
Staphylococcus not *aureus* (SSNA), for example, *S. epidermidis*	Cocci in clusters	Flat, white	Gamma	−	−	S
Staphylococcus saprophyticus	Cocci in clusters	Flat, white	Gamma	−	−	R
Staphylococcus lugdunensis	Cocci in clusters	Flat, white	Gamma	−	+	R

S, sensitive; *R*, resistant; −, negative; +, positive.

includes, but is not limited to, Group A streptococci (GAS), Group B streptococci (GBS), and Group C streptococci (GCS) as described below. The Mitis–Sanguinis group includes the species *Streptococcus pneumoniae*, while the Mutans group includes the species *Streptococcus mutans*. The Salivarius group includes *Streptococcus salivarius* and the Anginosus group includes the *Streptococcus anginosus*. The Bovis group includes *Streptococcus gallolyticus* and a final miscellaneous group includes other bacteria not described in detail here. The enterococci are described here to continue inclusion as the classical Group D streptococci.

Group A Streptococci *(Streptococcus pyogenes)*

GAS can be part of the normal flora or can be acquired through direct contact with respiratory secretions. Signs and symptoms of pharyngitis present 2–4 days after exposure with abrupt onset of sore throat, fever, malaise, and headache. The posterior pharynx may be erythematous with exudate with possible cervical lymphadenopathy [15]. Skin infections can also occur in the form of impetigo, cellulitis, erysipelas, and puerperal sepsis. Systemic infections can occur 1–2 days after pharyngitis symptoms begin. Scarlet fever begins with a diffuse erythematous rash on the chest and spreads to the extremities. The rash does not develop around the mouth, soles, or palms. A yellow/white coating develops on the tongue, and is eventually shed resulting in a red and raw surface, known as "strawberry tongue." This systemic infection is associated with strains containing the lysogenic phage which produces toxin. TSS and necrotizing fasciitis can also result from GAS infection.

Poststreptococcal sequelae include acute rheumatic fever (ARF) and acute glomerulonephritis (AGN). ARF follows pharyngitis and causes myocarditis, arthritis and arthralgia, inflammatory changes in multiple joints, blood vessels, subcutaneous tissues, chorea, subcutaneous nodules, and erythema marginatum. Rheumatic strains have M-protein epitopes similar to heart muscle, which results in molecular mimicry. AGN follows pharyngitis and/or pyoderma and impetigo. Inflammation occurs in renal glomeruli with edema, hypertension, hematuria, and proteinuria. Nephrogenic strains have proteins that bind to glomerulus resulting in antibody–antigen deposition [16]. GAS can be differentiated from the other β streptococci by their susceptibility to bacitracin.

Group B Streptococci *(Streptococcus agalactiae)*

Transient carriage in the gastrointestinal and genitourinary tracts is common. Neonatal diseases include bacteremia, pneumonia, and meningitis. In pregnant women, wound infection, urinary tract infections, and postpartum endometritis can occur. In adults who are elderly or have debilitating underlying conditions disease includes sepsis, pneumonia, bone and joint infections, skin and soft tissue infections [17]. Transmission can occur from mother to child before or during childbirth. Screening is performed at 35–39 weeks of pregnancy using a vaginal/rectal swab that is incubated in selective media such as Lim broth overnight and then sub-cultured for any colonies of β streptococci. The CAMP test can be used to differentiate GBS from β streptococci due to the production of CAMP factor. A streak of *S. aureus* is made onto a blood agar plate along with a perpendicular streak of isolate in question, not allowing the streaks to touch. An area of increased hemolysis in the form of an arrowhead where the organisms intersect due to release of CAMP factor by the GBS, and the beta-hemolysin secreted by *S. aureus*.

Group C Streptococci *(Streptococcus equi and Streptococcus dysgalactiae)*

These streptococci can be normal inhabitants of the skin and mucosal membranes. In immunocompromised patients, these streptococci can cause pharyngitis, bacteremia, and endocarditis. No specific test is available to specifically identify this organism that will be resistant to bacitracin and CAMP negative. Lancefield typing using latex agglutination can specifically differentiate Group A, B, C, and G streptococci.

Group D Streptococci

Enterococcus spp. (*Enterococcus faecalis* and *Enterococcus faecium*) (previously [18] *Streptococcus*) are part of the normal flora of the GI tract and can cause infection when introduced to other sites, commonly urinary, biliary tract, wound infections, bacteremia, endocarditis, intraabdominal, and pelvic infections. Enterococci commonly cause nosocomial infections, especially with long-term use of broad spectrum antibiotics (due to endemic resistance to cephalosporins) [18]. *Enterococcus gallolyticus* (previously known as *Streptococcus bovis*) can cause bacteremia and is associated with colon cancer. They appear as Gram-positive cocci in pairs and short chains on Gram stain. Colony morphology is small, gray, and γ-hemolytic. *E. faecalis* and *E. faecium* are bile esculin positive, and can grow in 6.5% NaCl. They are also positive for PYR. *E. gallolyticus* can be differentiated because although it is bile esculin positive, there is no growth on 6.5% NaCl, and is PYR negative. *Enterococcus cassileflavus* and *Enterococcus gallinarum* usually have a yellow pigment and demonstrate low level resistance to vancomycin.

Streptococcus pneumoniae

S. pneumoniae can be normal microbial flora in the nasopharynx of healthy individuals. Disease occurs when the organism is spread to other areas and associated infections can occur particularly in patients without a spleen. Other disease includes community-acquired pneumonia, sinusitis and otitis media, bacteremia, and meningitis [19]. Transmission occurs via respiratory droplets. The colony morphology is flat to mucoid α-hemolytic colonies on blood agar. The Gram stain reveals distinct Gram-positive lancet-shaped cocci in pairs. Colony morphology is α hemolytic colony often mucoid appearing like a donut. Biochemical tests reveal bile solubility and are sensitive to optochin, using P disks. This differentiates *S. pneumoniae* from the other α streptococci or viridans streptococci. There is a vaccine produced from capsular polysaccharides and is useful especially for high-risk patients with heart or lung disease, inmates, or asplenic and a second vaccine more useful in children.

Mutans Group (e.g., *Streptococcus mutans*)

The mutans group contains species that are normal flora in the upper respiratory and gastrointestinal tracts. A common infection with *S. mutans* is dental caries. Bacteremia and endocarditis are also possible. Colony morphology is flat, α-hemolytic colonies on blood agar plates.

Salivarius Group (e.g., *Streptococcus salivarius*)

S. salivarius is part of the normal flora of the oral cavity in humans and can enter the bloodstream transiently, especially after brushing teeth. Bacteremia and endocarditis are rare occurrences but can be seen in immunocompromised patients as a subacute infection.

Anginosus Group (e.g., *Streptococcus anginosis*)

S. anginosus is part of the normal microbiota but can cause bacteremia. The source of bacteremia is often abscess formation which can occur in the brain, oropharynx, or peritoneal cavity [19]. This group is β-hemolytic, and can include the Group antigens A, C, F, or G.

Catalase Negative Gram-Positive Cocci

Abiotrophia and *Granulicatella* (previously called nutritionally deficient streptococci) can cause bacteremia and endocarditis, but are also normal flora of the upper respiratory tract [20]. These fastidious organisms require pyridoxal or L-cysteine for growth. Satellite colonies can grow around *S. aureus* streaked on a BAP. Catalase negative cocci are summarized in Table 6.2.

Vancomycin Resistant Gram-Positive Cocci

Leuconostoc and *Pediococcus* cause opportunistic infections, including bacteremia, wound and CNS infections, endocarditis, and peritonitis [21,22]. They appear as α hemolytic colony on BAP and can resemble streptococci. Biochemical tests are used to differentiate these organisms from other Gram-positive cocci. PYR tests are negative and the ability to produce leucine aminopeptidase and hydrolyze esculin (ESC) can differentiate the two. Other Gram-positive cocci are summarized in Table 6.3.

GRAM-POSITIVE BACILLI

Gram-positive bacilli appear as rod shaped organisms and may have spores if they are depleted of nutrients on some media. Gram-positive bacilli will grow on blood agar plates and some that produce pigmentation such as *Rhodococcus* may appear as irregular rods. Some genera will stain positive with a weak acid fast stain such as *Nocardia* and appear as filaments in the Gram stain.

Aerobic Gram Positive Bacilli

Gram stain morphology

Spore forming catalase positive *Bacillus* sp
- Egg yolk positive
  - Motile β-hemolytic → *B. cereus*
  - Non-motile → *B. anthracis*
- Egg yolk negative → *Bacillus* sp

Nonspore forming
- Catalase positive → *Corynebacterium*, *Listeria* β, *Propionibacterium*
- Catalase negative → *Erysipelothrix* H_2S+, *Lactobacillus* Van^R, *Arcanobacterium* β, *Gardnerella*, *Actinomyces*

Branching rods → *Nocardia*, *Streptomyces*, *Rhodococcus*, RGM

Pigmented rods → *Methylobacterium*, *Nocardia*, *Rhodococcus* (cocci)

Table 6.2 Catalase negative cocci

Organism	Gram stain morphology	Colony on BAP	Hemolysis	Pyrrolidonyl arylamidase	Bile Esculin	6.5% NaCl	Optochin (P-disk)	Bacitracin
Pyogenic grp., for example, *Streptococcus pyogenes* (Group A streptococci)	Cocci in chains	Large, flat, smooth	Large beta zone	+				R
Pyogenic grp., for example, *Streptococcus agalactiae* (Group B streptococci)	Cocci in chains	Large, flat, smooth	Small beta zone	–				
Mitis–Sanguinis grp., for example, *Streptococcus pneumoniae*	Lancet-shaped diplococci	Small, mucoid	Alpha	–	V		S	
Mutans grp., for example, *Streptococcus mutans*	Cocci in chains	Tiny, green	Alpha	–	+		R	
Salivarius grp., for example, *Streptococcus salivarius*	Cocci in chains	Tiny, green	Alpha	–	V			
Anginosus grp., for example, *Sanginosis anginosis*	Cocci in chains	Tiny, green	Beta	–	+			
Bovis grp., for example, *Streptococcus gallolyticus*	Cocci in chains	Tiny, green	Gamma	–	V	No growth		
Enterococcus spp.	Cocci in pairs/chains	Small, gray, smooth	Gamma	+	+	Growth		

S, sensitive; *R*, resistant; –, negative; +, positive.

Table 6.3 Other Gram-positive cocci

Organism	Gram stain morphology	Colony	Hemolysis	Vancomycin	Catalase	Pyrrolidonyl arylamidase	Leucine amino-peptidase	Esculin
Abiotrophia	Gram-negative, coccobacilli	Alpha	Small, gray-white	S	–	+	w+	–
Granulicatella	Gram-negative, coccobacilli	Alpha	Tiny, gray	S	–	+	w+	–
Leuconostoc	Gram-positive coccobacilli	Alpha	Small, smooth gray	R	–	–	–	V
Pediococcus	Gram-positive, cocci	Alpha	Small, gray	R	–	–	+	+

S, sensitive; *R*, resistant; *w*, weakly; –, negative; +, positive.

Gram Positive, Endospore-Forming Bacilli

Most aerobic endospore-forming bacilli are ubiquitous, environmental organisms and do not normally cause disease. Endospores produced are highly resistant to extreme heat, desiccation, and radiation and can therefore survive in any environment including hospitals, food products, and in farms.

Bacillus spp.

Bacillus cereus is an environmental organism that may cause opportunistic infections. Local and systemic infections as well as food poisoning can occur in immunocompromised patients [23]. Infection from insufficient cleaning of hospital laundry and devices has been reported [24]. *Bacillus anthracis* is a selective agent as it can cause infection (anthrax) through aerosols. Infections can also occur via traumatic implantation as for the other *Bacillus* sp. Laboratory-acquired infections can occur and laboratory technicians should perform limited testing under the hood and send the isolate to the city health department for confirmation. Differentiation can be made between *B. cereus* and *B. anthracis* from the other *Bacillus* sp. by production of lecithinase which produces an opaque zone around the colonies on egg yolk agar. *B. cereus* is beta-hemolytic on blood agar plates, while *B. anthracis* is nonhemolytic.

Aerobic Gram Positive, Nonendospore-Forming bacilli

Some Gram-positive bacilli do not produce an endospore and are commonly found as part of the normal microbiota and in the environment.

Listeria

Listeria monocytogenes is found in the environment and has been isolated from various animals. *L. monocytogenes* can be found in processed meats, fresh produce, and milk products [25]. The bacteria can survive in cold temperatures and has been found in sandwich meats and ice cream. Pregnant women are asked not to eat foods that have been linked to *Listeria* outbreak due to the ability of the bacteria to cross the placenta and infect the fetus. Infection can lead to abortion, still birth, or preterm labor [26]. The organism can also cause meningitis in elderly patients. They appear as tiny β-hemolytic colonies on BAP and resemble B hemolytic streptococci. Differentiation would be made by appearance of rods versus cocci on Gram stain and production of catalase. *Listeria* is also motile particularly at room temperature and appears as an umbrella in motility agar. Penicillin and ampicillin can be used for treatment, *Listeria* is resistant to cephalosporins.

Corynebacterium

The *Corynebacterium* family has lesser than 100 species. Species that cause infection include *Corynebacterium diphtheriae*, of which the toxigenic form can cause throat and cutaneous diphtheria and the nontoxigenic form can cause endocarditis and pharyngitis. Toxigenic strains can be identified using the Elek test which is an immunodiffusion test using antibody to the diphtheria toxin. Immunization for this bacterium has almost removed the disease from countries with access to the vaccine. *Corynebacterium urealyticum* has been found to cause urinary tract infections, while *Corynebacterium jeikeium*, which is commonly multi-drug resistant (MDR) is associated with artificial devices and has been found to be the cause of endocarditis in such patients [27]. *Corynebacterium* appear as nonhemolytic or γ-hemolytic colonies on BAP. They are small Gram-positive rods that appear in clumps or as Chinese letters.

Lactobacillus

Lactobacilli species are found as normal microbiota of the gastrointestinal tract and vaginal flora and have historically been thought to be a beneficial bacterium in protection from bacterial vaginosis. However, bacteremia and endocarditis have been described with *Lactobacillus rhamnosus* [28]. In addition, the ingestion of probiotics containing *Lactobacillus* in immunocompromised patients has been shown to be a predisposing factor to infection with the same bacteria [29]. The use of probiotics continues to incite discussion as research in understanding the gut microbiome continues to grow [30]. The organism appears as α hemolytic colony on BAP and can be differentiated from streptococci by appearing as long chaining rods on Gram stain versus cocci in chains. They are also commonly resistant to vancomycin.

Erysipelothrix spp.

Erysipelothrix spp. are environmental organisms associated with animals. Infection occurs due to traumatic implantation. The organisms appear as small Gram-positive rods on Gram stain and small γ-hemolytic colonies on BAP. They are catalase negative and produce H_2S on TSI slants.

Weakly Acid Fast Bacteria

Aerobic actinomycetes contain mycolic acids in their cell walls and will stain better with an acid fast stain than with a Gram stain.

Nocardia spp.

Nocardia spp. are opportunistic soil organisms that can cause disease in immunocompromised patients or through traumatic implantation. Various species are noted to cause infection in humans [31]. *Nocardia asteroides* complex are thought to be the most common cause of nocardiosis and includes many species of varying susceptibility including *Nocardia abscessus* which has been shown to cause pulmonary and brain infections, *Nocardia nova* and *Nocardia farcinica*. *Nocardia brasiliensis* causes mycetoma including cellulitis and abscesses. *Nocardia* has been shown to cause respiratory infections. These bacteria will appear as long branching, beaded, Gram-positive bacteria, and weakly acid fast positive. Species differentiation can be accomplished by use of xanthine, tyrosine, and casein hydrolysis media. *N. asteroides* does not hydrolyze any of the substrates, while *N. brasiliensis* hydrolyzes casein and tyrosine forming a white precipitate around the colony. Antimicrobial therapy varies depending on the species and isolates should be tested for sensitivity.

Streptomyces

The most common infections caused by *Streptomyces* spp. are mycetoma. This group of bacteria contains species that are used commercially and can be found to colonize without causing disease [32]. The bacteria will appear as filamentous branching rods and will stain Gram positive in a Gram stain; however, they will not stain with acid fast stain differentiating the organism from *Nocardia*.

Rhodococcus spp.

Rhodococcus equi is the most commonly isolated pathogen. Infections caused by this organism are seen in immunocompromised patients and usually present as pulmonary infection, similar to tuberculosis with granuloma formation. Susceptibility testing should be performed, especially in the case of rifampin as some strains carry an *rpoB* mutation which confers rifampin resistance [33]. *Rhodococci* will appear as Gram-positive coccobacilli and can stain with a beaded appearance. They produce a peachy color on agar media after extended incubation.

Mycobacteria spp.

Mycobacteria spp. are nonmotile, nonspore forming aerobic rods. The cell wall contains mycolic acids stain and stain acid fast positive which appear pinkish/red with the acid fast stains.

Mycobacterium leprae

Mycobacterium leprae causes granulomatous disease, anesthetic skin lesions, and nerve damage immune reactions. Spectrum of disease runs from tuberculoid leprosy with few bacilli and a granulomatous response to many organisms with little granulomatous response as lepromatous leprosy. The organisms do not grow in culture and are diagnosed by biopsy and evidence of organisms on Fite stain (weak acid fast) and the consistent pathologic response. Transmission can occur through shedding from the nose. Treatment can include dapsone, rifampicin, and clofazimine which are available only from the World Health Organization free of cost.

Mycobacterium tuberculosis

Mycobacterium tuberculosis complex comprises *M. tuberculosis*, *M. bovis*, *M. bovis* BCG, *Mycobacterium africanum*, among others. The organisms divide every 12–24 hours, requiring weeks before growth is detected and are niacin and nitrate positive. Pulmonary infection is a slow progression of chronic inflammation and caseation with the subsequent formation of cavities. When the foci rupture, there are a large number of organisms that spread to the lungs and are aerosolized. Transmission occurs via these aerosolized droplets from coughing of infected person.

Decontamination of the specimen is required to release mycobacteria trapped in cells and concentration of specimen is recommended to recover mycobacteria, as well as to remove other normal flora. Direct detection of *M. tuberculosis* complex in respiratory specimens and identification of the presence of the *rpoB* gene that determines rifampicin resistance can be made using molecular techniques. Two different interferon-gamma release assays are whole blood tests that measure a person's immune reactivity to *M. tuberculosis*. Both T-Spot and QuantiFERON can aid in diagnosis *M. tuberculosis* but do not differentiate latent infection from tuberculosis disease.

An important virulence factor is that of the cord factor, a glycolipid called trehalose 6,6'-dimycolate or TDM, is present in the outer envelope of all mycobacteria and protects the bacteria from the host response [34].

Organisms in the *M. tuberculosis* complex are buff colored on solid media, have a cording appearance when grown in broth and cannot be differentiated from each other by DNA probe or PCR. However, *M. tuberculosis* is positive for both niacin and nitrite and *M. bovis* is negative. MTB is also NAP test positive, pyrazinaimidase positive, and grows on media with TCH, while *M. bovis* is negative for NAP and pyrazinimidase (resistant to PZA) and will not grow on media with TCH. Treatment for

M. tuberculosis includes an extended regimen with pyrazinamide, isoniazid, ethambutol, and rifampicin (if not resistant).

Nontuberculous Mycobacteria

There are currently about 150 species of nontuberculous mycobacteria, about half of which are slow growers. These bacteria are normally found in the environment and occasionally cause disease, especially in immunocompromised patients. Pigmentation of nontuberculous mycobacteria can be used to classify various species into nonchromogenic or chromogenic. Photochromogenic species will produce a pigment only after exposure to light. Scotochromogenic mycobacteria will produce pigment when grown in the dark or after exposure to light. The classical grouping for NTM is called Runyon grouping:

Runyon I: Photochromogens produce little or no pigment when grown in the dark but become highly pigmented when grown in light.

Runyon II: Scotochromogens are slow growing, and produce a yellow-orange pigment in light or in the dark. Some become darker with exposure to light.

Runyon III: Nonchromogens are nonpigmented in the light and dark or have only a pale yellow, buff, or tan pigment that does not intensify after light exposure.

Runyon IV: Rapid growers will grow on media within 7 days.

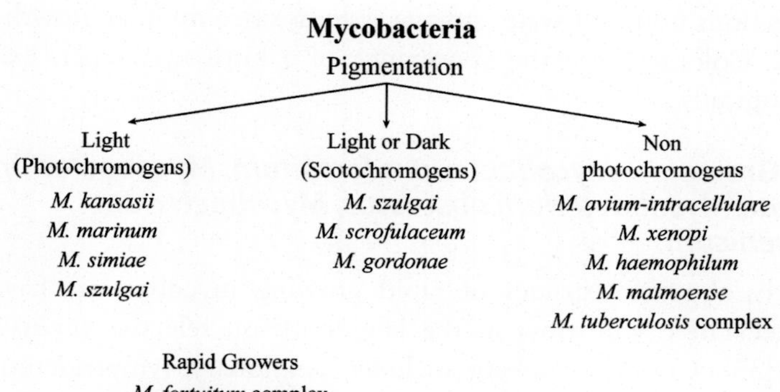

Photochromogens

Mycobacterium kansasii causes chronic pulmonary infection that resembles pulmonary tuberculosis particularly in adult males exposed to asbestos or other chemicals that might have caused lung damage. The organism also causes cervical lymphadenitis in children and disseminated as well as cutaneous infections in immunocompromised patients. *M. kansasii* can be found in tap water. They have a yellow pigment when grown on Lowenstein Jensen media in presence of light. Treatment includes rifampin and clarithromycin.

Scotochromogens

Mycobacterium szulgai can cause chronic pulmonary disease similar to tuberculosis. *M. szulgai* is considered scotochromogenic at 37°C and photochromogenic at 25°C. *Mycobacterium gordonae* infection is rare and pathogenicity of this species is low. *M. gordonae* is a casual resident in human sputum and gastric lavage specimens. The colonies have an orange pigment when grown on solid media.

Nonphotochromogens

Mycobacterium avium complex (*M. avium* and *Mycobacterium intracellulare*) can cause pulmonary disease and may also be present as a disseminated disease. In patients with AIDS, *M. avium* is the most common disease causing environmental nontuberculous mycobacteria. *M. avium* can be present as disseminated disease. Treatment includes macrolides, rifampicin, and ethambutol.

Mycobacterium xenopi can be found on skin lesions of a cold blooded animal, *Xenopus laevus. M. xenopi* can cause hospital-acquired infections and infections from hot water storage systems since optimal growth occurs at 42°C. Colonies have the appearance of a birds nest when examined microscopically.

Rapid Growers—*Mycobacterium fortuitum, Mycobacterium chelonae/Mycobacterium abscessus, Mycobacterium mucogenicum*

Currently, about 70 species of rapid growing mycobacteria have been identified. These will grow in the laboratory on selective media within 7 days [35]. Common disease includes pulmonary, lymphadenitis, skin, and soft tissue infections. These organisms grow within 7 days after subculture on solid media and act more like routine bacteria. They will stain as Gram-positive bacilli on Gram stain. They are arylsulfatase positive and grow on a modified version of MacConkey agar and this differentiates them from most other mycobacteria species. Treatment includes clarithromycin, azithromycin, cephalosporins, carbapenems, and quinolones and depends on susceptibility testing of the isolate.

Skin Mycobacteria—*Mycobacterium marinum, Mycobacterium haemophilum*, and *Mycobacterium ulcerans*

The most common skin mycobacteria causing human infections is *Mycobacterium marinum* which is associated with freshwater wound infections. These species require growth at 30°C as compared to 35°C for routine mycobacteria species. In addition, *Mycobacterium haemophilum* require supplementation with hemin. *Mycobacterium ulcerans* is the cause of Buruli ulcer which mainly occurs in Africa, parts of S. America, and Australia. Biochemical differentiations of mycobacteria are summarized in Table 6.4.

GRAM-NEGATIVE BACTERIA

Gram-negative bacteria are ubiquitous and reside in the intestines as normal flora in humans. These Gram negatives are facultative anaerobes and can be differentiated based on their ability to ferment glucose, reduce nitrate, and produce catalase or oxidase. Most are motile and contain flagella. In the laboratory, most Gram negatives will grow on blood agar media, and MacConkey agar will select for these Gram negatives and differentiate the lactose fermenters from nonfermenters. Gram negatives have various virulence factors including capsules that protect the bacteria from phagocytosis. The cell wall contains lipopolysaccharide, which is made of three components, the O polysaccharide, core polysaccharide, and lipid A. Lipid A triggers fever, vasodilation, inflammation, shock, and disseminated intravascular coagulation [36].

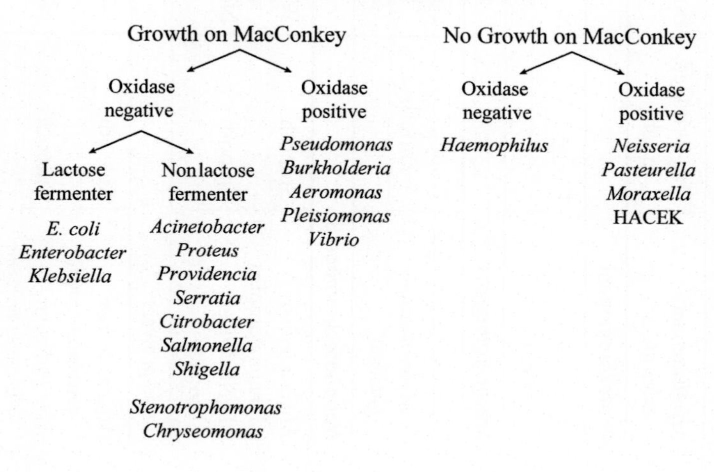

Table 6.4 Mycobacterial biochemical differentiation

Organism	Löwenstein–Jensen slant	Pigmentation	Growth	Heat stable catalase 68°C	Arylsulfatase	Urease	Nitrate reduction	Niacin	Tellurite reduction
Mycobacterium tuberculosis	Dry, wrinkled, rough	Buff	2–4 weeks	+/−	−	+	+	+	−
Photochromogens									
Mycobacterium kansasii	Rough	Yellow	10–20 days	+	−	+	+	−	−
Scotochromogens									
Mycobacterium szulgai, Mycobacterium gordonae	Rough/smooth	Orange	7–14 days	+	−	−	+	−	+
Nonphotochromogens									
Mycobacterium avium complex	Smooth	White, yellow	2–4 weeks	+	−	−	−	−	+
Mycobacterium xenopi	Smooth	None, white		+	+/−	−	−	−	−
Rapid growers									
Mycobacterium fortuitum complex	Smooth	None, white	2–7 days	+	+	+	+	−	+
Mycobacterium chelonae/abscessus complex	Smooth	None, white	2–7 days	V	+	+	−	−	+

−, negative; +, positive; *V*, variable.

Enterobacteriaceae are negative for oxidase production and ferment glucose. Members of the family can be differentiated by their ability to ferment lactose. These bacteria are normally found in the gastrointestinal tract of humans and cause disease when they are present in other areas.

Lactose Fermenters

There are several bacteria which are lactose fermenters. They are discussed in this section.

Escherichia coli

Escherichia coli can cause extraintestinal and diarrheagenic disease in humans. Extraintestinal disease includes urinary tract infections and meningitis. Diarrheagenic disease has multiple categories that include bacteria which cause bloody diarrhea, hemorrhagic colitis, and enteroinvasive disease [37]. Transmission can occur through ingestion of contaminated food and water, person to person contact, and contact with animals or fomites.

E. coli will grow on MacConkey agar plates and are positive for lactose fermentation. The colonies appear as pink, flat, and dry. To identify *E. coli* O157:H7 (enterohemorrhagic strains, EHEC) or serotypes that can be associated with hemolytic uremic syndrome, sorbitol MacConkey agar can be used. The O157:H7 serotype cannot ferment sorbitol and appear as clear colonies while the other *E. coli* ferment sorbitol and appear pink as on plain MacConkey agar. The other diarrheagenic *E. coli*, enteropathogenic or enterotoxigenic, are indistinguishable by growth on agar media and therefore require molecular techniques for differentiation. *E. coli* can also biochemically resemble *Shigella*. However, *E. coli* is usually motile, while *Shigella* is not. Supportive care is recommended for diarrheagenic infection and antibiotic therapy is not beneficial [38]. For extraintestinal disease, ampicillin, cefazolin, gentamicin, and tobramycin should be evaluated for susceptibility. There are an increasing number of bacteria which produce extended spectrum beta-lactamases (ESBLs) in addition to carbapenemases.

Klebsiella pneumoniae

Klebsiella pneumoniae is a common cause of hospital-acquired pneumonia and common cause of infection in immunocompromised patients [39]. Humans can be carriers in the nasopharynx and bowel, where fecal contamination is the most common cause of infections. *K. pneumoniae* is a lactose fermenters on MacConkey and appears mucoid due to the polysaccharide capsule that will confer resistance to phagocytosis [40]. There are an increasing number of bacteria which produce ESBLs in addition to carbapenemases, therefore susceptibility testing is recommended. *K. pneumoniae* is intrinsically resistant to ampicillin.

Slow/Nonlactose Fermenters

There are several bacteria which are nonlactose fermenters. They are discussed in this section.

Citrobacter freundii and Citrobacter koseri

Citrobacter freundii and *Citrobacter koseri* can cause urinary tract infections, and are found in wound, respiratory, meningitis, and sepsis. They can cause healthcare-associated infections, especially in pediatric and immunocompromised patients [41]. Third-generation cephalosporins are the choice of treatment for meningitis. *Citrobacter* spp. are intrinsically resistant to ampicillin. *C. freundii* is intrinsically resistant to ampicillin-sulbactam and cephalosporins. They can resemble *Salmonella* on agar media due to production of H_2S by both genera. They are also citrate positive, however, only *Citrobacter* is positive for ONPG.

Serratia

Serratia spp. are opportunistic organisms and can cause healthcare-associated infections. *Serratia* spp. can produce beta-lactamase and possess ESBLs. *Serratia* is intrinsically resistant to ampicillin, ampicillin-sulbactam, cephalosporins, nitrofurantoin, and polymyxin antibiotics. Many isolates will produce a red pigment when grown on agar media.

Salmonella

Salmonella spp. can cause various types of disease. Typhoid fever causes bloodstream infection in underdeveloped countries. Nontyphoidal diseases can cause intestinal infections that last 7 or more days. *Salmonella* is transmitted through the ingestion of contaminated water and food including poultry and eggs. Additionally, animal contact with reptiles and amphibians has been shown to cause outbreaks [42]. The organism can remain in the gall bladder and be shed intermittently. When grown on Hektoen enteric agar, *Salmonella* colonies will appear as blackish colonies due to the production of hydrogen sulfide in the presence of thiosulfate.

Antibiotic therapy is not recommended for gastroenteritis. For invasive and typhoidal *Salmonella* infections, antibiotic susceptibility testing is recommended, though fluoroquinolones and third-generation cephalosporins are reasonable empiric treatment [43].

Shigella

Shigella can cause dysentery, bloody diarrhea, and sometimes bloodstream infections. Transmission usually occurs from person to person, especially child care centers and through ingestion of contaminated food and water is common. They are not transmitted from animal contact as for *Salmonella*. When grown on Hektoen enteric agar colonies will appear as green colonies and are distinguishable from *Salmonella* because they do not produce hydrogen sulfide in the presence of thiosulfate. The infectious dose is low and can cause disease with very few cells. Disease can be due to the production of Shiga toxin and be associated with HUS as for *E. coli* EHEC. It is negative for most biochemical except for methyl red. Antibiotic therapy is not required since shigellosis is self-limiting; however, antibiotics have been shown to decrease duration of fever and diarrhea [44].

Yersinia spp.

Yersinia species are found as normal flora in many animal species. Humans are usually infected from consumption of infected animal products or from the environment. They are unusual in that they can survive cold temperatures. This can be used as a selection method to allow growth from contaminated food products. This also allows the organisms to survive in red blood cells and cause transfusion-related infections. *Yersinia enterocolitica* can present clinically as appendicitis due to growth in the ileum. *Yersinia pseudotuberculosis* can produce necrotizing granulomas and resemble tuberculosis. *Yersinia pestis* is the causative agent of plague and is considered as an agent of bioterrorism. Plague is acquired from animal, particularly rats.

Proteus, Providencia, and Morganella

Proteus, *Providencia*, and *Morganella* are organisms that are part of the normal microbiota of the gastrointestinal tract. They do not ferment lactose; produce H_2S and *Proteus mirabilis* grows as a swarming colony on BAP. They are most commonly associated with urinary tract infections due to the formation of urease and production of stones.

Nonenterobacteriaceae

Acinetobacter

Acinetobacter baumannii complex has been shown to cause ventilator-associated pneumonia and bloodstream infections. *Acinetobacter lwoffii* is a known skin colonizer but has been found to cause bacteremia in hospitalized patients. *Acinetobacter* are coccobacilli and can resemble *Neisseria* species by Gram stain. Growth on MacConkey will appear as nonfermenters and do not produce oxidase. *A. baumannii* have a purple appearing colony. Due to the high occurrence of developing resistance, a panel of antibiotics should be tested. For MDR strains, colistin can be tested and utilized.

Stenotrophomonas

Stenotrophomonas maltophilia is an opportunistic pathogen that has been shown to cause variety of diseases including meningitis, urinary tract infection, and wound infections [45]. Growth on MacConkey will appear as nonfermenters and are negative for oxidase production. *S. maltophilia* is intrinsically resistant to beta lactams, aminoglycosides, and quinolones as well as carbapenems. Trimethoprim-sulfamethoxazole is the drug of choice for treatment.

Neisseria spp.

Neisseria gonorrhoeae causes gonorrhea and often causes acute urethritis in men with urethral discharge. *N. gonorrhoeae* is transmitted through sexual contact and infects the mucosal surfaces and is always a pathogen. Asymptomatic infections can occur in 10% of cases. Females are mostly affected in the endocervix and sometimes the urethra. The infection may ascend and result in pelvic inflammatory disease. Asymptomatic infection in females is common. *Neisseria meningitidis* can be either a commensal organism or a pathogen. Infection can be present as meningitis and sepsis. Septic shock can have lethality of 30%. *N. meningitidis* is considered normal flora of the oropharynx, but can cause acute disease in healthy patients especially when found in sterile body fluids.

N. gonorrhoeae can be grown on chocolate agar, however use of selective media, such as Thayer–Martin which contains the antibiotics vancomycin and colistin for the suppression of Gram-negative and Gram-positive bacteria is necessary for growth from cervical specimens. Due to the fastidious nature of the organism and the need to plate specimens directly on media at patient's bedside many laboratories have eliminated the use of culture for genital specimens and have opted to use molecular methods for detection. *N. meningitidis* can grow on blood agar and chocolate plates collected from CSF, blood, or joint fluid. On Gram stain the organisms appear as Gram-negative diplococci. Colonies suspicious for

Neisseria species can be confirmed by oxidase (positive) and by carbohydrate utilization. *N. gonorrhoeae* is glucose positive, *N. meningitides* is glucose and maltose positive, and *Neisseria lactamica* is glucose, maltose, and lactose positive. If all sugars are negative and the organisms appears as a *Neisseria* species it is likely *Moraxella*. *N. gonorrhoeae* can be treated with ceftriaxone however strains with increased resistance are of concern. *N. meningitides is treated with penicillin or ceftriaxone*. Contacts are prophylactically given rifampin or a quinolone. Vaccines are available as a quadrivalent polysaccharide–protein conjugate vaccine conserving serogroups A, C, W, and Y. Vaccine is given to college age students and offered to laboratory workers due to the ease with which it can be contracted.

GRAM-NEGATIVE BACILLI

Various Gram-negative bacilli are discussed in this section.

Pseudomonas aeruginosa

Immunocompromised patients have a higher risk of wound and mucosal surface infections. It can cause burn wound infections, urinary tract infections, folliculitis, and is the predominant respiratory tract pathogen in patients with cystic fibrosis (CF) [46]. Colonies grow as nonlactose fermenters on MacConkey. They are β-hemolytic and can produce pyocyanin pigment which gives them a dark appearance on BAP and a greenish pigment on clear media such as Muellar–Hinton agar. They can also be specifically identified by their ability to grow at 42°C. MDR strains of *Pseudomonas aeruginosa* exist, especially in CF patients, and antibiotic susceptibility testing is recommended.

Burkholderia spp.

Burkholderia spp. causes a variety of disease. *Burkholderia cepacia* complex causes disease in patients with CF especially pneumonia, bacteremia, UTIs, and septic arthritis. *Burkholderia pseudomallei* is considered as a bioterrorism agent and causes melioidosis. *Burkholderia mallei* is also considered as a selective agent and causes glanders. Colonies can grow on MacConkey agar and selective media such as PC agar. On a Gram stain, the Gram-negative bacilli will stain in a bipolar manner, resembling safety pins. *B. cepacia* complex are intrinsically resistant to aminoglycoside and polymyxin and can be resistant to beta-lactam antibiotics. *B. pseudomallei* are intrinsically resistant to penicillin, aminoglycosides, and macrolides. Infections can usually be treated with ceftazidime, carbapenems, doxycycline, and trimethoprim-sulfamethoxazole.

Vibrio

Vibrio cholera is divided into many serogroups, of which O1 and O139 cause cholera. Outbreaks have occurred in Haiti, the Americas, and Africa. The organisms can cause diarrhea, cellulitis, and septicemia. *Vibrio parahaemolyticus* is associated with eating raw fish and shellfish found in Asia and the United States. Infections can cause watery diarrhea, bacteremia, and wound infections. *Vibrio vulnificus* causes septicemia and wound infections with over 90% of cases caused after eating raw oysters or traumatic implantation of the organism in patients with underlying liver disease.

The organisms can grow on MacConkey, Hektoen enteric, or Salmonella-Shigella, agar and will appear as colorless colonies. TCBS agar is selective for vibrio in which sucrose can allow differentiation of *V. cholera* which can ferment sucrose and form yellow colonies, from *V. parahaemolyticus* and *V. vulnificus* which do not ferment sucrose and produce green colonies. *V. cholera* are also string test positive. For noncholera vibrio, antibiotic susceptibility testing can be performed on ampicillin and tetracycline.

Aeromonas

Aeromonas spp. are Gram-negative coccobacilli that cause acute, dysentery, and acute gastroenteritis. Symptoms can include abdominal pain, fever, nausea, and vomiting. Transmission of *Aeromonas* will occur in aquatic environmental areas, brackish water. Colonies grow on MacConkey agar and species are beta-hemolytic on blood agar plates.

Plesiomonas shigelloides

The only oxidase positive member of the Enterobacteriaceae, formerly known as *Aeromonas*, is the major cause of traveler's diarrhea in Japan and China. Three types of gastroenteritis can occur including secretory, invasive, and subacute/acute. *Plesiomonas* is found in freshwater, raw, and undercooked shellfish. Colonies appear as nonlactose fermenters on MacConkey. Supportive therapy and oral rehydration is recommended for treatment, and only severe cases of infection should receive antimicrobial therapy, namely ciprofloxacin.

Fastidious Gram-Negative Rods

Various fastidious Gram-negative rods are discussed in this section.

Haemophilus spp.

Haemophilus spp. do not grow on MacConkey agar. *Haemophilus influenzae* has been shown to cause invasive infection such as meningitis and there is direct spread from water droplets from the upper respiratory tract of infected individuals. The development of the *H. influenza* type B vaccine has decreased mortality. *Haemophilus ducreyi* causes chancroid which is a sexually transmitted disease that leads to a single genital ulcer. *H. ducreyi* usually occurs in Asia, Africa, and Latin America. *Haemophilus parainfluenzae* can be a colonizer of the upper respiratory tract and can cause sinusitis and bronchitis. *H. influenzae* do not grow on BAP and grow only on chocolate agar as they require X and V factors for growth. *H. parainfluenza* require only V factor and *H. ducreyi* only requires X factor. *Haemophilus* spp. do not produce oxidase. Production of beta-lactamase production should be tested for isolates. Cephalosporins and carbapenems can be used as well as combination agents with a beta-lactamase inhibitor.

HACEK Organisms: *Aggregatibacter spp., Cardiobacterium spp., Eikenella corrodens,* and *Kingella spp.*

These bacteria are part of the normal flora and transmission can occur through wounds and interaction with the oral cavity of humans or animals that are colonized. *Haemophilus aphrophilus* is now *Aggregatibacter aphrophilus*. *Aggregatibacter actinomycetemcomitans* can cause periodontitis and *Cardiobacterium* mainly causes endocarditis. *Eikenella corrodens* can cause periodontitis, infections in the upper respiratory tract, joints, and wounds, especially human bites. *Kingella* spp. can cause bone and joint infections, and septic arthritis, especially in pediatric patients. Production of beta-lactamase should be tested, while third-generation cephalosporins should be considered drug of choice.

Campylobacter

Campylobacter causes gastrointestinal infection and diarrheal illness. *Campylobacter* is a zoonotic pathogen and can contaminate foods such as chickens, cattle, sheep, and pigs. On Gram stain organisms appear as seagull, shaped Gram-negative rods. Colonies will grow on *Campylobacter* selective medium such as Campy-CVA (cefoperazone, vancomycin, and amphotericin) or Skirrow media and require extra concentrations of CO_2 for growth and a temperature of 42°C. *Campylobacter* spp. are oxidase positive. *Campylobacter jejuni* can be differentiated from *Campylobacter coli* by hydrolysis of hippurate. Infection can be treated with azithromycin and erythromycin.

Helicobacter

Helicobacter causes diarrheal illness and is associated with peptic ulcer disease and is a risk factor for gastric cancer. *Helicobacter* is part of the gastrointestinal tract of many mammalian animals and while mode of transmission is still controversial, though there is evidence for oral to oral and fecal to oral transmission. The organisms, which used to be classified as *Campylobacter*, are Gram-negative spiral bacteria. Diagnosis is commonly made using the urea breath test (organisms produce urease) or stool antigen. Stool specimens and gastric biopsy specimens can be cultured; however, the organisms are quite fastidious and slow growing. Treatment requires a proton pump inhibitor, clarithromycin, and amoxicillin and/or metronidazole. Culture is useful in cases of treatment failure so that antimicrobial susceptibility can be performed on the isolate.

Bordetella

Bordetella pertussis causes pertussis, also known as whooping cough. There are two phases of the catarrhal phase which consists of sneezing and coughs and the paroxysmal phase which consists of coughing spasms, whooping, and vomiting. *Bordetella parapertussis* causes milder disease with a shorter duration. Transmission occurs via air droplets with a low infectious dose. These fastidious organisms are slow growing and require selective media such as Regan Lowe or Bordet-Gengeau. Collection of nasopharyngeal swabs for molecular testing is the preferred method of diagnosis. Treatment for infection includes penicillin, macrolides, tetracycline, and trimethoprim/sulfamethoxazole.

Bartonella spp.

Bartonella henselae causes cat-scratch disease and presents systemic infection and with possible skin lesions. Lymphadenitis develops 2–3 weeks after exposure and can persist for several weeks. *Bartonella quintana* causes trench fever which is a recurrent fever that causes 3–5 febrile episodes. *B. henselae* and *B. quintana* can cause bacillary angiomatosis which is a vasculoproliferative disease of the skin. Transmission occurs through insect vectors, fleas, or ticks, while cats are the main reservoir. Diagnosis is usually clinical since they are difficult to grow in culture. When cultured for extended periods of incubation colonies are tiny and usually negative for most biochemical reactions including catalase and oxidase. Treatment for infection is doxycycline.

Brucella spp.

Brucellosis presents with nonspecific signs and symptoms including fever, weight loss, and loss of appetite. Transmission of *Brucella* is through ingestion of unpasteurized products from various animals depending on the species (cattle *Brucella abortus*, goats and sheep *Brucella melitensis*, and pigs *Brucella suis*). The organism is most commonly recovered from culture of blood or bone marrow. Blood cultures may need to be incubated for longer than the usual 5 days of incubation to recover *Brucella* spp. Tiny colonies may be apparent on BAP or chocolate agar after 24–28 hours of incubation. Care should be taken for all blood cultures which required longer than usual time for growth and whose Gram stain demonstrated poorly stained tiny Gram-negative coccobacilli. These organisms should be handled with minimal biochemical tests including positive oxidase, catalase, and urease. As soon as culture is suspicious for *Brucella* the isolate should be sent to a reference laboratory in the Laboratory Response Network for confirmation of identification. *Brucella* spp. are the most common cause of laboratory-acquired infections. Infection is treated with doxycycline and rifampin.

Francisella tularensis

Francisella tularensis subspecies *tularensis* (type A) found in North America and *F. tularensis* subspecies *holarctica* (type B), with wider distribution in the Northern Hemisphere, are the causative agents of tularemia. Disease is nonspecific including chills, fever, headache, and malaise. The various forms include oculoglandular, glandular, oropharyngeal, typhoidal, and pneumonic, with the most common form, ulceroglandular. *F. tularensis* is highly contagious and can be ingested, inhaled, or can be carried via transfer of an arthropod vector. Infection can also be acquired by contact with infected animals while skinning them. Colonies will appear on chocolate and BCYE media after 48 hours. Gram stain will show tiny Gram-negative coccobacilli. They are oxidase negative, weakly positive catalase, beta-lactamase positive, and urease negative. If *F. tularensis* cannot be ruled out, then the cultures must be sent to the Sentinel laboratory

Table 6.5 Bacteria and their vectors

Organism	Disease	Reservoir	Vector	Vector common name	Geographical location
Rickettsia rickettsii	Rocky Mountain Spotted Fever	Mammal	*Rhipicephalus/ Dermacentor/ Amblyomma*	Brown dog tick/ American Levi tick/Lone Star tick	United States, Mexico, Central and South America
Rickettsia akari	Rickettsialpox	Mice	*Allodermanyssus*	Mouse mite	NYC, Europe, Ukraine
Rickettsia typhi	Endemic murine typhus	Rat	*Xenopsylla cheopis*	Rodent flea	Southern California, Texas
Rickettsia prowazekii	Epidemic typhus	Humans	*Pediculus humanus*	Head and body lice	Worldwide
Orientia tsutsugamushi	Scrub typhus	Rodents	*Trombicula*	Chigger	Asia-Pacific region Russia and China to Indonesia, North Australia to Afghanistan
Ehrlichia chafeensis	Ehrlichosis	Rodents	*Amblyomma*	Lone Star tick	United States
Anaplasma phagocytophilum	Anaplasmosis	Small mammals, deer	*Ixodes*	Deer tick	Northeast, Upper Midwestern United States
Francisella tularensis/ Francisella holarctica	Tularemia	Small mammals	*Amblyomma*	Lone Star tick	North America/Europe and Asia
Borrelia burgdorferi	Lyme disease	Small mammals, birds	*Ixodes*	Deer tick	North America Europe
Yersinia pestis	Bubonic plague	Rodents	*Xenopsylla cheopis*	Rodent flea	Rare/worldwide
Coxiella	Q fever	Mammals	*Rhipicephalus/ Dermacentor/ Amblyomma*	Brown dog tick/ American Levi tick/Lone Star tick	United States

for confirmation. Treatment for infection includes streptomycin and gentamicin. Table 6.5 contains a list of organism with their respective vectors and hosts.

Pasteurella multocida

Pasteurella multocida is most commonly associated with infections in cattle and rabbits. Infections in humans are usually acquired following cat bites. The organisms are also fastidious, appear as tiny Gram-negative coccobacilli, and will grow on BAP or chocolate after several days. They are catalase, oxidase, and indole positive.

Legionella

Legionella pneumophila is an intracellular organism and the causative agent of Legionnaire's disease. This name comes from an epidemic pneumonia which occurred after a convention of the Pennsylvania State Legionnaires in Philadelphia, PA, in 1976 [47]. The bacteria can reside in the environment, especially in warm water supplies. Transmission occurs via air droplets and outbreaks have been seen in community cooling towers or other contaminated potable water systems. *L. pneumophila* replicates within macrophages and monocytes and causes a pneumonia which can have a range of mild to fatal disease especially with patients who have risk factors such as smoking, heart, or lung disease and immunocompromised patients. Pontiac fever is an acute, self-limiting, and flu-like disease that has been associated with exposure to *Legionella*-type environments. Also fastidious small Gram-negative coccobacilli, isolation can be difficult on BCYE with 0.1% α-ketoglutaric acid and requires decontamination to reduce contamination of normal flora.

Capnocytophagia

Capnocytophagia spp. can cause septicemia and endogenous infections. Disease has been reported after a patient had a recent dog or cat bite. Because these are fastidious organisms, they will only grow on chocolate and BAP and require CO_2 for growth. Colonies of the bacteria are very small and will look yellow when scraped off the plate. The colonies appear to spread due to their gliding motility.

Streptobacillus moniliformis

Streptobacillus moniliformis is the causative agent of rat bite fever, a systemic illness that can range from rash on the extremities to septicemia. Isolation of the organism is best made from body fluids on BAP and requires CO_2 for growth.

Tropheryma whipplei

Tropheryma whippelei is the causative agent of Whipple's disease and found in the environment. The organism is likely ingested in early childhood and rarely causes disease. Whipple's disease causes chronic diarrhea and arthralgia of large joints which can eventually lead to a picture of wasting syndrome. The organism has not ever been cultured so diagnosis is made by typical histologic morphology of a lymph node with foamy histiocytes full of organisms or by molecular techniques.

ANAEROBES

These organisms grow better in the absence of oxygen and strict anaerobes will not grow at all in the presence of oxygen. They are normal flora of the mouth, upper respiratory tract, and GI and GU tracts. Most can be presumptively identified by their microscopic morphology on a Gram stain, growth on selective agars, and a few biochemical tests in addition to susceptibility to antimicrobial agents by disk testing.

Anaerobic Cocci

Peptostreptococcus are anaerobic Gram-positive cocci which appear in chains in a Gram stain. The cocci are much smaller and slower growing than the aerobic Gram-positive cocci. *Veilonella* is anaerobic Gram-negative cocci which can look like *Neisseria* species on Gram stain.

Anaerobic Gram Positive, Endospore-Forming Bacilli

Anaerobic Gram-positive bacilli can be differentiated by the ability to produce spores. Nonspore forming bacilli can then be separated by a catalase and indole test.

Clostridium spp.

Clostridium spp. cause a wide array of infections depending on the species. *Clostridium septicum* is known to cause bloodstream infections. *Clostridium tetani* is the cause of tetanus, terminal spores (tennis racket), *Clostridium botulinum* can cause foodborne illness, *Clostridium perfringens* causes gas gangrene, and *Clostridium difficile* is the cause of the colitis. Spores can be found in soil, feces, and sewage. Disease of the *Clostridia* can be due to the extensive number of toxins produced by each species. More than 25 toxins are known to cause disease in the various clostridium, including eurotoxins, enterotoxins, collagenases, permeases, neurotoxin, toxin, lipases, lecithinases, hemolysin, proteinases, hyaluronidases, DNases, neuraminidase, botulinum exotoxin, and tetanus [48]. Growth of the bacteria in the laboratory on media is difficult and testing for the bacteria has become easier with molecular and serological testing. On Gram stain they appear as large box car shaped rods, similar in appearance to *Bacillus* spp. One exception is *C. perfringens* which can be grown from wound infections. On BAP the organisms has a characteristic double zone of hemolysis and is positive for lecithinase on EYA, which differentiates them from other *Clostridium* spp.

C. difficile, cause of antibiotic associate diarrhea or pseudomembranous colitis, has two toxins, A and B, that are the targets for molecular testing. An enzyme immunoassay (EIA) can used to distinguish toxigenic bacteria. An EIA for glutamate dehydrogenase is an enzyme that is produced by all strains of *C. difficile*, and must be used in combination with the toxin EIA to determine if the strain produces the toxin. Treatment for *C. difficile* is metronidazole and vancomycin.

Propionibacterium

Propionibacterium is a common normal flora of the skin especially of the head, neck, upper back, and chest and a cause of acne. The organism can cause infection following surgical procedures to those areas of the body or in patients with artificial devices such as shunts or rods especially when located in the upper back or shoulder. The organism is slow growing, facultative anaerobe that can grow aerobically and is catalase positive. Incubation is usually extended to allow for growth of this organism in suspicious specimens.

Actinomyces spp.

Actinomyces israelli is a strict anaerobe that is normal vaginal flora and mouth flora. It can be associated with infections of IUDs and infections of the jaw which produce sulfur granules. The organism is very slender Gram-positive bacilli that typically branch. They grow on blood agar plates after anaerobic incubation as a white, molar tooth colony and grow in an anaerobic broth as bread crumbs.

Gram Negative, Nonspore Forming Bacilli

Anaerobic Gram-negative bacilli can be identified based on morphology. *Bacteroides* spp. are small Gram-negative rods that grow as black colonies on *Bacteroides* bile esculin agar due to ESC hydrolysis, while *Prevotella* and *Porphyromonas* do not. Presumptive identification can be made by susceptibility to colistin, vancomycin, and kanamycin using antimicrobial disks. Antimicrobials that these organisms are commonly used to treat infections due to these organisms include metronidazole and carbapenems and due to the production of β-lactamase, a β-lactams with a β-lactamase inhibitor such as piperacillin/tazobactam.

Anaerobic Gram-Negative Bacilli

Gram stain morphology

Small GNR

Pleomorphic GNR
Va^r KM^s CL^s

BBE Agar

Pigmentation

Fusobacterium
Indole

Black Col
Va^r KM^r CL^r

Negative

Positive

+

=

VA^r

VA^s
Indole +

F. nucleatum
(Pointed ends)

F. necrophorum

B. fragilis

Prevotella

Porphyromonas

Bacteroides fragilis

Bacteroides fragilis are common colonizers of the gastrointestinal tract, mucosal surfaces, and oral cavities of animals and humans. Spread of the organisms to adjacent tissues and into bloodstream can cause infection. They can cause acute appendicitis, bacteremia, endocarditis, and intraabdominal abscesses. *Bacteroides* will appear as black colonies on *Bacteroides* bile esculin (BBE) agar, and are resistant to kanamycin, vancomycin, and colistin.

Prevotella

Prevotella spp. is a common colonizer of the oral microbiota. Infections include aspiration pneumonia, lung abscess, pulmonary embolism, and bacteremia. *Prevotella* spp. appear as greenish colonies on BBE agar and can be differentiated from *Bacteroides* by the lack of tolerance to bile and hydrolysis of ESC which causes the black colonies of *Bacteroides*. Some species can produce brick red fluorescence.

Fusobacterium

Fusobacterium spp. are common colonizers of the mouth microbiota. They can be associated with abscesses, periodontal disease, skin ulcers, and

Lemierre's syndrome. On anaerobic media, the bacteria appear greenish on BBE agar and are distinguishable from *Prevotella* by its sensitivity to colistin. *Fusobacterium nucleatum* appear as long thin Gram-negative bacilli with pointed ends and are indole positive.

Chlamydia spp.

Chlamydia is obligate intracellular bacteria that can only grow in cell culture as do viruses and not on solid media. They have two forms the elementary body and the reticulate body (infectious form). *Chlamydia* causes a wide range of infections depending on the species. *C. trachomatis* contains various serovars which can cause a large range of disease including follicular keratoconjunctivitis causing blindness, genital tract infection which causes nongonococcal urethritis in men and *cervicitis* in women and lymphogranuloma venereum (LGV). *Chlamydophila pneumoniae* causes upper and lower respiratory tract infections. *Chlamydophila psittaci* can cause a severe chronic pneumonia with mild symptoms including fever, chills, and severe headaches. The natural reservoir are birds and can be is transmitted via inhalation of fecal material or direct contact of infected birds.

Antibodies to the outer membrane protein can be used for diagnosis after growth in tissue culture or on slides prepared from scrapings containing the organism. However, the diagnosis is typically made by molecular detection of the organisms in either specimen from the genital or respiratory tract depending on the species suspected. Diagnosis of LGV is made by serology by detection of antibodies to the LGV serotypes of *C. trachomatis*.

Rickettsia spp. *and Orientia*

Rickettsia spp. are intracellular organisms that are transmitted by various types of arthropods. The species are subdivided into the spotted fever group which has multiple species (including *Rickettsia rickettsii* and *Rickettsia akari*) and the typhus group (*Rickettsia prowazekii* and *Rickettsia typhi*). Pathogenic species of *Rickettsia* are maintained in their respective animal and arthropod reservoir and are accidentally transmitted by arthropod vectors to human hosts. The most common human pathogen is *R. rickettsii*, the causative agent of Rocky Mountain spotted fever. *R. akari* is the causative agent of rickettsialpox and has been reported in cosmopolitan areas where it is maintained in the rodent population. *R. powazekii* is the cause of epidemic, or louse-borne typhus. Epidemic typhus occurs in areas with people living in crowded, unsanitary conditions that can arise during wars, famines, and natural disasters. *R. typhi* is the cause of endemic or murine typhus. Endemic disease is distributed in warm, humid areas including the Gulf States and southern California.

Orientia tsutsugamushi formally belonged to the *Rickettsia* genus and is the causative agent of scrub typhus. Symptoms of infection include fever, macular rash, and lymphadenopathy.

These organisms are not able to grow in culture and diagnosis is made by serology. Table 6.5 contains a list of organism with their respective vectors and hosts.

Ehrlichia and *Anaplasma*

Ehrlichia spp. are intracellular organisms that are transmitted to vertebrates by ticks and is the causative agent of ehrlichiosis. *Ehrlichia chaffeensis* and *Ehrlichia ewingiii* are the two species that can cause human disease. *E. chaffeensis* is the cause of human monocytic ehrlichiosis which presents headache, malaise, myalgia, and nausea. CNS involvement and rashes can occur especially in children.

Anaplasma spp. is the causative agent of human granulocytic anaplasmosis and is transmitted via tick bites. Disease includes high fever, myalgia, malaise, and headaches. They may be seen on a peripheral blood smear using a Giemsa or Wright–Giemsa stain, particularly in HGE.

Coxiella burnetii

Coxiella burnetii is an intracellular pathogen and the causative agent of Q fever. Q fever includes acute, febrile illness, as a chronic infection that may involve the cardiovascular system, or as a postinfection chronic fatigue syndrome. Table 6.5 contains a list of organism with their respective vectors and hosts. Diagnosis is made by serology.

Mycoplasma and *Ureaplasma*

Mycoplasma spp. do not have cells walls and do not stain with a Gram stain. These bacteria require enriched growth medium with nucleic acids. *Mycoplasma hominis* appear as "fried-egg" colonies while *Mycoplasma pneumoniae* appear as spherical colonies. *M. pneumoniae* has been associated with atypical pneumonia in children and young adults which can be associated with autoimmune hemolytic anemia due to the production of cold agglutinins which is an IgM antibody. Disease presents tracheobronchitis and upper respiratory tract infections. *M. pneumoniae* is difficult to grow in culture and therefore diagnosis is made by serology or PCR. *Mycoplasma genitalium* causes urethritis, cervicitis, endometritis, and conjunctivitis. *M. hominis* can cause pyelonephritis, cervicitis, bacteremia, and pneumonia.

Ureaplasma urealyticum causes urethritis and is transmitted via sexual contact in humans.

SPIROCHETES

The spirochetes are bacteria which have a corkscrew appearance and are difficult to culture in the laboratory. *Treponema* and *Borrelia* are obligate intracellular organisms. Serology and antigen testing is the most reliable method to diagnose disease.

Treponema pallidum

Treponema pallidum is the causative agent of syphilis. Primary syphilis presents hard, painless, and broad-based chancre. Secondary syphilis presents a symmetrical widely distributed macular rash. The rash can infect mucosal membranes and may appear on the palms and soles. Secondary syphilis occurs weeks after the lesion has healed and is considered the most infective stage of disease. A latent period follows secondary syphilis before tertiary syphilis presents with personality changes, blindness, and gummas. Darkfield microscopy can be used to rapidly diagnose syphilis with a wet preparation of a skin or lesion scraping. However, this assay has poor sensitivity and the CDC reverse algorithm is recommended for the diagnosis of syphilis. This includes the use of an EIA assay which is specific for *T. pallidum* and once positive lasts for life. If positive a nonspecific RPR is performed in order to get a titer. This titer typically decreases over time and can be used to gage the need for treatment. Diagnosis of CNS disease is typically made clinically or by using the nonspecific VDRL test on CSF, although this test has poor sensitivity.

Borrelia burgdorferi

Borrelia burgdorferi is the causative agent of Lyme disease and is transmitted to humans by ticks. Infection is characterized by a "bull's eye" skin lesion, which develops at the site of the tick bite. Symptoms can include fever, fatigue, headache, arthralgia, and myalgia. Table 6.5 contains a list of organism with their respective vectors and hosts.

Leptospira interrogans

Leptospira interrogans are motile, spiral shaped organisms with hooked ends, and are found in the environment in association with the urine of animals. They can cause Weil disease which presents meningitis, hepatitis, and nephritis. These spirochetes can grow on Fletcher's semisolid medium however it can take up to 8 weeks of incubation and requires the performance of a darkfield microscopic exam of the medium to visualize growth of the organisms. Diagnosis is therefore made by serology.

SELECT AGENTS

The Laboratory Response Network links local laboratories with state and federal laboratories to assist in the case of public health emergencies. Select agents are those organisms that have the potential to cause mass harm to people due to the small inoculum that causes disease and ability to spread via aerosols. *Francisella*, *Burkholderia*, and *Brucella* will grow slowly and will take 72 hours for mature colonies to appear. These organisms will grow initially on blood agar but poorly or not at all upon subculture. If the characteristics in Table 6.6 are seen in an isolate in the clinical laboratory, then the select agents cannot be ruled out and must be sent to the reference laboratory for confirmatory testing.

KEY POINTS

- Bacteria are among the most common organisms to cause disease. Clinical treatment requires an understanding of basic physical characteristics of each organism, as well as an appreciation of epidemiology and pathophysiology.
- In the laboratory, biochemical assays allow differentiation between both genera and species, and supports decision trees for antimicrobial therapy. Initial examinations include Gram stains for bacteria, followed by inoculation in aerobic and anaerobic media to allow for growth of the organisms.
- Gram-positive bacteria include the *Staphylococcus* species characterized by a thin polysaccharide capsule composed primarily of monosaccharides and proteins useful for cell attachment and biofilm formation. These bacteria represent common skin colonizers. Coagulase positive *Staphylococcus* can cause both cutaneous and systemic infection. The majority of isolates are resistant to penicillin due to the production

Table 6.6 Common rule out tests of select agents

Organism	Gram stain result	Media	Catalase	Oxidase	Urease	Indole
Bacillus anthracis	Gram-positive bacilli	BAP, Chocolate	+	–	–	
Francisella tularensis	Gram-negative coccobacillus	BAP, Chocolate	weakly +	V/+		–
Burkholderia mallei and *Burkholderia pseudomallei*	Gram-negative coccobacillus	BAP, Chocolate	+	V/+		–
Brucella spp.	Gram-negative coccobacillus	BAP, Chocolate	+	+	+	
Yersinia pestis	Gram-negative rod	BAP, Chocolate, MacConkey	+	–	–	–

V, variable; +, positive; –, negative.

of β-lactamase. The MRSA represent an especially important clinical group.

- The streptococci differ from the staphylococci in regards to their lack of catalase activity. They are facultative anaerobes and appear as Gram-positive cocci in pairs and/or chains on a Gram stain. These bacteria can be characterized by the varying hemolytic processes. The Streptococci are classified by Lancefield grouping serological analysis. Streptococci are all sensitive to penicillin. Other clinically important Gram-positive cocci include those that are catalase negative as well as those that are vancomycin resistant.
- Aerobic Gram-positive bacilli appear as rod shaped organisms and grow on blood agar plates. Some produce pigmentation that readily permits clinical identification. The aerobic endospore-forming bacilli do not normally cause disease unless the infected host is immunocompromised. The aerobic Gram positive, nonendospore-forming bacilli can usually be identified based on characteristics of appearance of rods versus cocci on Gram stain, and through their production of catalase. Anaerobic Gram-positive bacilli can be differentiated by the ability to produce spores.
- Many organisms may be initially differentiated by acid fast staining, with detection based on presence of mycolic acids in cell wall structure.
- Gram-negative bacteria are facultative anaerobes and can be differentiated based on their ability to ferment glucose, reduce nitrate, and produce catalase or oxidase. Some contain flagella. Differentiation may be made in the laboratory by growth on selective agar, detection of virulence factors, or presence of fermentation pathways.
- The Enterobacteriaceae are negative for oxidase production and ferment glucose. Members of the family can be differentiated by their ability to ferment lactose. These bacteria are normally found in the gastrointestinal tract. The clinically important nonenterobacteriaceae usually are classified as nonfermenters.
- Oxidase positive bacteria represent a unique subset of organisms that cause a variety of disease states. Clinical recognition takes into account production of pigments, growth on selective agar, and susceptibility to specific antimicrobial agents.
- Fastidious Gram-negative bacilli can cause invasive infections, such as meningitis. The HACEK organisms (*Aggregatibacter* spp., *Cardiobacterium*

spp., *E. corrodens*, and *Kingella* spp.) are especially important in periodontitis and endocarditis, as well as upper respiratory tract infections.

- The spirochetes are bacteria which have a corkscrew appearance and are difficult to culture in the laboratory.
- Anaerobic Gram-negative bacilli can be identified by morphology. Identification can be made by susceptibility to colistin, vancomycin, and kanamycin using antimicrobial disks.
- The *Mycobacteria* spp. are nonmotile, nonspore forming aerobic rods with cell walls that contain unique mycolic acids which are preferential for acid fast stains. Many cause granulomatous disease; the nontuberculous mycobacteria only cause disease in immunocompromised hosts. Speed of growth in culture can also assist in determination of species.

REFERENCES

[1] Relman DA. The human body as microbial observatory. Nat Genet 2002;30:131–3.

[2] Woese CR. Microbiology in transition. Proc Natl Acad Sci USA 1994;91:1601–3.

[3] Marshall JH, Wilmoth GJ. Pigments of *Staphylococcus aureus*, a series of triterpenoid carotenoids. J Bacteriol 1981;147:900–13.

[4] Thammavongsa V, Kim HK, Missiakas D, Schneewind O. Staphylococcal manipulation of host immune responses. Nat Rev Microbiol 2015;13:529–43.

[5] Balaban N, Rasooly A. Staphylococcal enterotoxins. Int J Food Microbiol 2000;61:1–10.

[6] Mulligan ME, Murray-Leisure KA, Ribner BS, Standiford HC, John JF, Korvick JA, et al. Methicillin-resistant *Staphylococcus aureus*: a consensus review of the microbiology, pathogenesis, and epidemiology with implications for prevention and management. Am J Med 1993;94:313–28.

[7] Spaulding AR, Salgado-Pabon W, Kohler PL, Horswill AR, Leung DY, Schlievert PM. Staphylococcal and streptococcal superantigen exotoxins. Clin Microbiol Rev 2013;26:422–47.

[8] Rogolsky M. Nonenteric toxins of *Staphylococcus aureus*. Microbiol Rev 1979;43:320–60.

[9] Vandenesch F, Naimi T, Enright MC, Lina G, Nimmo GR, Heffernan H, et al. Community-acquired methicillin-resistant *Staphylococcus aureus* carrying Panton-Valentine leukocidin genes: worldwide emergence. Emerg Infect Dis 2003;9:978–84.

[10] Hamory BH, Parisi JT, Hutton JP. *Staphylococcus epidermidis*: a significant nosocomial pathogen. Am J Infect Control 1987;15:59–74.

[11] Buttner H, Mack D, Rohde H. Structural basis of *Staphylococcus epidermidis* biofilm formation: mechanisms and molecular interactions. Front Cell Infect Microbiol 2015;5:14.

[12] Hovelius B, Mardh PA. *Staphylococcus saprophyticus* as a common cause of urinary tract infections. Rev Infect Dis 1984;6:328–37.

[13] Huang CR, Lu CH, Wu JJ, Chang HW, Chien CC, Lei CB, et al. Coagulase-negative staphylococcal meningitis in adults: clinical characteristics and therapeutic outcomes. Infection 2005;33:56–60.

[14] Janda W. The genus *Streptococcus*, vol. 36. New York, NY: Elsevier; 2014.

[15] Walker MJ, Barnett TC, McArthur JD, Cole JN, Gillen CM, Henningham A, et al. Disease manifestations and pathogenic mechanisms of Group A Streptococcus. Clin Microbiol Rev 2014;27:264–301.

[16] Martin WJ, Steer AC, Smeesters PR, Keeble J, Inouye M, Carapetis J, et al. Post-infectious Group A streptococcal autoimmune syndromes and the heart. Autoimmun Rev 2015;14:710–25.
[17] Schuchat A. Epidemiology of Group B streptococcal disease in the United States: shifting paradigms. Clin Microbiol Rev 1998;11:497–513.
[18] Sava IG, Heikens E, Huebner J. Pathogenesis and immunity in enterococcal infections. Clin Microbiol Infect 2010;16:533–40.
[19] Chao Y, Marks LR, Pettigrew MM, Hakansson AP. Streptococcus pneumoniae biofilm formation and dispersion during colonization and disease. Front Cell Infect Microbiol 2014;4:194.
[20] Christensen JJ, Facklam RR. *Granulicatella* and *Abiotrophia* species from human clinical specimens. J Clin Microbiol 2001;39:3520–3.
[21] Starr JA. *Leuconostoc* species-associated endocarditis. Pharmacotherapy 2007;27:766–70.
[22] Iwen PC, Mindru C, Kalil AC, Florescu DF. Pediococcus acidilactici endocarditis successfully treated with daptomycin. J Clin Microbiol 2012;50:1106–8.
[23] Schoeni JL, Wong AC. *Bacillus cereus* food poisoning and its toxins. J Food Prot 2005;68:636–48.
[24] Sifuentes LY, Gerba CP, Weart I, Engelbrecht K, Koenig DW. Microbial contamination of hospital reusable cleaning towels. Am J Infect Control 2013;41:912–5.
[25] Lomonaco S, Nucera D, Filipello V. The evolution and epidemiology of *Listeria monocytogenes* in Europe and the United States. Infect Genet Evol 2015;35:172–83.
[26] Poulsen KP, Czuprynski CJ. Pathogenesis of listeriosis during pregnancy. Anim Health Res Rev 2013;14:30–9.
[27] Bernard K. The genus corynebacterium and other medically relevant coryneform-like bacteria. J Clin Microbiol 2012;50:3152–8.
[28] Gouriet F, Million M, Henri M, Fournier PE, Raoult D. *Lactobacillus rhamnosus* bacteremia: an emerging clinical entity. Eur J Clin Microbiol Infect Dis 2012;31:2469–80.
[29] Vanichanan J, Chavez V, Wanger A, De Golovine AM, Vigil KJ. Carbapenem-resistant Lactobacillus intra-abdominal infection in a renal transplant recipient with a history of probiotic consumption. Infection 2016;44:793–6.
[30] Sanchez B, Delgado S, Blanco-Miguez A, Lourenco A, Gueimonde M, Margolles A. Probiotics, gut microbiota, and their influence on host health and disease. Mol Nutr Food Res 2017;61.
[31] Wilson JW. Nocardiosis: updates and clinical overview. Mayo Clin Proc 2012;87:403–7.
[32] Joseph NM, Harish BN, Sistla S, Thappa DM, Parija SC. Streptomyces bacteremia in a patient with actinomycotic mycetoma. J Infect Dev Ctries 2010;4:249–52.
[33] Fines M, Pronost S, Maillard K, Taouji S, Leclercq R. Characterization of mutations in the rpoB gene associated with rifampin resistance in *Rhodococcus equi* isolated from foals. J Clin Microbiol 2001;39:2784–7.
[34] Hunter RL, Olsen M, Jagannath C, Actor JK. Trehalose 6,6'-dimycolate and lipid in the pathogenesis of caseating granulomas of tuberculosis in mice. Am J Pathol 2006;168:1249–61.
[35] Brown-Elliott BA, Wallace Jr. RJ. Clinical and taxonomic status of pathogenic non-pigmented or late-pigmenting rapidly growing mycobacteria. Clin Microbiol Rev 2002;15:716–46.
[36] Annane D, Bellissant E, Cavaillon JM. Septic shock. Lancet 2005;365:63–78.
[37] Kaper JB, Nataro JP, Mobley HL. Pathogenic *Escherichia coli*. Nat Rev Microbiol 2004;2:123–40.
[38] Thielman NM, Guerrant RL. Clinical practice. Acute infectious diarrhea. N Engl J Med 2004;350:38–47.
[39] Podschun R, Ullmann U. *Klebsiella* spp. as nosocomial pathogens: epidemiology, taxonomy, typing methods, and pathogenicity factors. Clin Microbiol Rev 1998;11: 589–603.

[40] Domenico P, Salo RJ, Cross AS, Cunha BA. Polysaccharide capsule-mediated resistance to opsonophagocytosis in *Klebsiella pneumoniae*. Infect Immun 1994;62:4495–9.
[41] Samonis G, Karageorgopoulos DE, Kofteridis DP, Matthaiou DK, Sidiropoulou V, Maraki S, et al. Citrobacter infections in a general hospital: characteristics and outcomes. Eur J Clin Microbiol Infect Dis 2009;28:61–8.
[42] Centers for Disease C, Prevention Reptile-associated salmonellosis-selected states, 1998-2002. MMWR Morb Mortal Wkly Rep 2003;52:1206–9.
[43] Humphries RM, Fang FC, Aarestrup FM, Hindler JA. In vitro susceptibility testing of fluoroquinolone activity against Salmonella: recent changes to CLSI standards. Clin Infect Dis 2012;55:1107–13.
[44] Christopher PR, David KV, John SM, Sankarapandian V. Antibiotic therapy for Shigella dysentery. Cochrane Database Syst Rev 2010. http://dx.doi.org/10.1002/14651858.CD006784.pub3:CD006784.
[45] Denton M, Kerr KG. Microbiological and clinical aspects of infection associated with *Stenotrophomonas maltophilia*. Clin Microbiol Rev 1998;11:57–80.
[46] Doring G, Flume P, Heijerman H, Elborn JS, Consensus Study Group. Treatment of lung infection in patients with cystic fibrosis: current and future strategies. J Cyst Fibros 2012;11:461–79.
[47] Fraser DW, Tsai TR, Orenstein W, Parkin WE, Beecham HJ, Sharrar RG, et al. Legionnaires' disease: description of an epidemic of pneumonia. N Engl J Med 1977;297:1189–97.
[48] Barth H, Aktories K, Popoff MR, Stiles BG. Binary bacterial toxins: biochemistry, biology, and applications of common Clostridium and Bacillus proteins. Microbiol Mol Biol Rev 2004;68:373–402, [table of contents].

CHAPTER 7

Antibiotics, Antimicrobial Resistance, Antibiotic Susceptibility Testing, and Therapeutic Drug Monitoring for Selected Drugs

Contents

Microbiology and Molecular Diagnosis in Pathology.
DOI: http://dx.doi.org/10.1016/B978-0-12-805351-5.00007-7

INTRODUCTION

Antibiotics represent a diverse group of chemotherapeutic agents with activity against microorganisms such as bacteria, fungi, viruses, or protozoa. The word antibiotic comes from Greek words "anti" means against and "bios" meaning life. Although the original word coined by Selman Waksman represented compounds which were derived from microorganisms, many antibiotics currently used in medical practice are also synthetic or semisynthetic compounds. The first effective antibiotic penicillin was discovered by Sir Alexander Fleming in 1928, but currently more than 100 antibiotics are approved for medical use worldwide. Antibiotics can be classified based on their chemical structures (Table 7.1). Alternatively, they can be categorized on the basis of their target specificity. The narrow-spectrum antibiotics target particular types of bacteria, such as

Table 7.1 Classification of antibiotics based on chemical structure

Chemical structure	Specific example of drugs
Aminoglycoside	Amikacin, gentamicin, kanamycin, neomycin, netilmicin, paromomycin, sisomicin, streptomycin, tobramycin
Glycopeptide	Vancomycin, teicoplanin
Beta-lactam (penicillin and related drugs)	Penicillin G, Penicillin V, ampicillin, carbenicillin, dicloxacillin, nafcillin, oxacillin, piperacillin, temocillin, ticarcillin
Beta-lactam (cephalosporins)	First generation: Cefadroxil, cefazolin, ceflatonin, cephalexin Second generation: Cefaclor, cefamandole, cefoxitin, cefprozil, cefuroxime Third generation: Cefixime, cefdinir, cefditoren, cefoperazone, cefotaxime, ceftriaxone, ceftizoxime Fourth generation: Cefepime Fifth generation: Ceftobiprole
Beta-lactam (carbapenem)	Ertapenem, doripenem, meropenem
Monobactam	Aztreonam, imipenem, meropenem, ertapenem
Macrolide	Azithromycin, clarithromycin, dirithromycin, erythromycin, roxithromycin, troleandomycin, spectinomycin
Polypeptides	Bacitracin, colistin, polymyxin B
Oxazolidinones	Linezolid, quinupristin/dalfopristin
Quinolones	Ciprofloxacin, enoxacin, gatifloxacin, moxifloxacin, ofloxacin, norfloxacin, levofloxacin
Sulfonamides	Mafenide, sulfacetamide, sulfadiazine, sulfamethoxazole, sulfanilamide, sulfisoxazole, trimethoprim
Tetracycline	Doxycycline, minocycline, oxytetracycline, tetracycline

Gram-negative or Gram-positive bacteria, while broad-spectrum antibiotics can be effective against a wide range of bacteria. In addition, antibiotics can be either bactericidal or bacteriostatic, based on their mechanism of action. Bactericidal agents typically kill bacteria directly, whereas bacteriostatic agents prevent cell growth and division [1,2].

Antibiotics can also be classified on the basis of their mechanism of action. There are six major mechanisms by which an antibiotic exerts its pharmacological action. These include:

- Inhibition of cell wall synthesis.
- Inhibition of bacterial protein synthesis.
- Disruption of the bacterial cell membrane.
- Damage to bacterial cell membrane.
- Inhibition of bacterial nucleic acid synthesis.
- Antimetabolite activities.

Inhibition of bacterial cell wall formation is probably the most common mechanism by which an antibiotic kills bacteria or inhibits bacterial growth. Antibiotics which interfere with cell wall synthesis are beta-lactam antibiotics (penicillin and related antibiotics), cephalosporins, vancomycin, etc., while antibiotic such as clindamycin, chloramphenicol, lincomycin, and macrolide antibiotics interferes with protein synthesis of bacteria by binding to the 50S ribosomal unit. Antibiotics that interfere with bacterial protein synthesis by binding to the 30S ribosomal unit are tetracycline and aminoglycosides. Sulfonamides and trimethoprim kills bacteria by inhibiting folate synthesis. Antibacterial effect of metronidazole, quinolones, and novobiocin is due to their capability of interfering with bacterial DNA synthesis while rifampin interferes with bacterial RNA synthesis. Polymyxin B kills bacteria by interfering with cell membrane function. In Table 7.2, antibiotics are classified based on their mechanism of action.

Table 7.2 Mechanism by which antibiotics kill bacteria

Mechanism	Examples of antibiotic
Inhibition of bacterial cell wall synthesis	Vancomycin, cephalosporins, beta-lactam antibiotics (including semisynthetic drugs)
Inhibition of bacterial protein synthesis	Aminoglycosides, chloramphenicol, macrolide antibiotics, tetracycline
Disrupting bacterial cell membrane	Bacitracin
Damaging bacterial cell membrane	Polymyxin B
Inhibition of bacterial nucleic acid synthesis	Quinolones, metronidazole, rifampin
Antimetabolite activity	Sulfonamides, dapsone, trimethoprim

INTRODUCTION TO PHARMACOKINETICS AND PHARMACODYNAMICS

When a drug is administered orally it undergoes several steps in the body that determine concentration of that drug in serum/plasma or whole blood. These steps include liberation (release of drug from binders) of the drug, absorption from gastrointestinal tract, distribution of the drug in target organ, protein binding of the drug, metabolism (hepatic or non-hepatic metabolism), and elimination. Although many drugs are excreted in the urine, some drugs are also excreted through bile. Basic pharmacokinetics has been discussed in detail in pharmacology textbooks. Therefore, only a brief discussion of this important topic is presented in this chapter. Liberation of a drug after oral administration depends on the formulation of the dosage. Immediate release formulation releases the drugs at once from the dosage form when administered, while the same drug may also be available in sustained release formulation. Absorption of a drug depends on the route of administration. Generally, an oral administration is the route of choice but certain drugs must be administered intravenously or intramuscularly if the drug has poor oral bioavailability, or is destroyed by gastric acid or undergoes extensive first pass metabolism. When a drug enters the blood circulation, it is distributed throughout the body into various tissues and the pharmacokinetic parameter is called volume of distribution (V_d). This is the hypothetical volume to account for all drugs in the body and is also termed as the apparent volume of distribution:

$$V_d = \text{Dose/Plasma concentration of drug}$$

The amount of a drug that interacts with the receptor or target site is usually a small fraction of the total drug administered. Drugs usually undergo chemical transformation (metabolism) before elimination. Many drugs are metabolized in the liver in two phases by various enzymes but cytochrome P-450 mixed function oxidase is the major liver enzyme responsible for metabolism of a majority of drugs in the Phase I step that involves manipulation of a functional group of a drug molecule in order to make the molecule more polar. The Phase II step may involve acetylation, sulfation, methylation, amino acid conjugation, or glucuronidation in order to increase polarity of the drug metabolite. The half-life of a drug is the time required for the serum concentration to be reduced by 50%. Therapeutic drug monitoring is recommended when a drug reaches a steady state. Half-life of a drug can be calculated from elimination rate constant (K) of a drug:

$$\text{Half-life} = 0.693/K$$

Elimination rate constant can be easily calculated from the serum concentrations of a drug at two different time points using the following formula where Ct_1 is the concentration of drug at a time point t_1 and Ct_2 is the concentration of the same drug at a later time point t_2:

$$K = \frac{\ln Ct_1 - \ln Ct_2}{t_2 - t_1}$$

Renal excretion is a major pathway for the elimination of drugs and their metabolites. Therefore, impaired renal function may cause accumulation of drugs and metabolites in serum thus increasing the risk of adverse drug effect. Moreover, other pathological conditions such as liver disease, congestive heart failure, and hypothyroidism may also decrease clearance of drugs. Drugs may also be excreted via other routes, such as biliary excretion. The factors which determine elimination of a drug through the biliary track include chemical structure, polarity, and molecular weight as well as active transport sites within the liver cell membranes for that particular drug.

BETA-LACTAMS AND CEPHALOSPORINS

Beta-lactam antibiotics penicillins and cephalosporins are still the most widely used group of antibiotics available, with penicillin being one of the oldest commercially available antibiotics. These drugs are part of a broad classes of antimicrobial agents known as beta-lactam antibiotics as they share a beta-lactam ring nucleus as part of their core molecular structure. These drugs are bactericidal and act to disrupt the synthesis of the peptidoglycan layer of bacterial cell walls and cell wall structural integrity. Beta-lactam antibiotics interfere with the final transpeptidation step and competitively inhibit crosslinking of peptidoglycan. All these antibiotics have wide therapeutic window and no therapeutic drug monitoring is needed during therapy with these antibiotics.

Penicillins, such as Penicillin G, Penicillin V, amoxicillin, nafcillin, and ampicillin, are used for treatment of Gram-positive infections such as *Staphylococcus* species, *Streptococcus pneumoniae*, beta-hemolytic strains of *Streptococcus*, and *Enterococcus faecalis*. These drugs may be used alone or in preparations that include beta-lactamase inhibitors, such as clavulanic acid, tazobactam, and sulbactam, for treatment of penicillin-resistant organisms.

Several generations (first through fifth) of cephalosporin antibiotics have now been developed and are widely prescribed with an increasingly broader spectrum of action against both Gram-negative and Gram-positive organisms. First-generation cephalosporins (e.g., cefazolin) are predominantly active against Gram-positive bacteria, and successive generations have increased activity against Gram-negative bacteria (albeit often with reduced activity against Gram-positive organisms). Later generations of cephalosporins, such as cefaclor, and ceftriaxone cefotaxime are effective against organisms including *Proteus mirabilis*, *Klebsiella* species, *Enterobacter aerogenes*, and *Haemophilus influenzae*, while ceftazidime and cefepime have activity against *Pseudomonas aeruginosa*. Sepsis and nosocomial infections are significant problems in intensive care units due to high mortality associated with such infections. Cefepime and ceftazidime are effective in treating Gram-negative nosocomial infections in critically ill patients. Available data indicates that cefepime may have some advantages over ceftazidime due to broader spectrum of activity and less potential for development of bacterial resistance [3].

Ceftaroline is a cephalosporin with expanded activity to include methicillin-resistant *Staphylococcus aureus* (MRSA) and vancomycin-resistant *S. aureus* (VRSA). The Gram-negative activity of ceftaroline includes common respiratory pathogens and members of the Enterobacteriaceae. Ceftaroline can also be used for treating community-acquired pneumonia and complicated skin infections [4]. Adverse drug reactions with beta-lactam antibiotics include diarrhea, nausea, rash, urticaria fever, vomiting, erythema, and dermatitis. Allergic hypersensitivity occurs in up to 10% patients, with anaphylaxis as a rare complication (most frequently with penicillins).

Mechanisms of Resistance

Beta-lactams and cephalosporins exert their antimicrobial effect by covalently binding with bacterial penicillin-binding proteins (PBPs) which are diverse enzymes (transpeptidases or carboxypeptidases) involved in bacterial cell wall synthesis and are anchored in the cytoplasmic membrane of the bacteria. When antibiotics bind to PBPs, bacterial cell wall synthesis is severely impaired due to cell wall lysis, disruption of cell shape, and inhibition of cell division which eventually causes cell death. However, for antibacterial action, these drugs must diffuse through the bacterial cell wall to bind with PBPs and the antibiotic must resist enzymatic degradation. Resistance to beta-lactam antibiotics may be due to alteration of PBPs with reduced binding with a specific antibiotic, decreased ability of the antibiotic to diffuse through bacterial cell wall, or bacteria may synthesize beta-lactamases enzymes that inactivate the antibiotic by hydrolysis of beta-lactam ring.

Resistance to beta-lactam antibiotics due to PBP modification is either related to mutations of genes encoding PBPs or through acquisition of supplementary new genes that encode new PBPs with low or no affinity to bind with beta-lactam antibiotics. This mechanism is seen more commonly in Gram-positive bacteria such as *S. pneumoniae* or *S. aureus* than Gram-negative bacteria. Bacteria are unicellular organisms and their cytoplasm is separated from the external environment by the cytoplasmic membrane. Gram-negative bacteria surround itself with an additional outer membrane which acts as a barrier for antibiotics to diffuse through the membrane with subsequent binding with PBPs. Usually antibiotics and nutrients diffuse through the protein channels on the cell membrane known as "porins." During antibiotic resistance, these porin channels are altered thus reducing diffusion of antibiotics in the bacterial cell. Alternatively bacteria may be resistant to a particular antibiotic due to alteration of specific porin channels responsible for diffusion of that drug. However, the most common mechanism of beta-lactam resistance is the production of beta-lactamases by resistant bacteria. Some beta-lactamases are encoded by chromosomal genes where others are encoded by plasmid genes [5]. Extended spectrum beta-lactamases (ESBLs) are a rapidly evolving group of beta-lactamases, commonly found in *E. coli*, *Klebsiella pneumoniae*, and *P. mirabilis*, which are capable of hydrolyzing third generation cephalosporins (ceftriaxone, ceftazidime, and cefotaxime) and aztreonam but are inhibited by clavulanic acid. The ESBLs are frequently encoded by plasmid genes and these enzymes are responsible for extreme drug resistance of some bacteria. Antibiotic options in the treatment of ESBLs producing organisms are limited, although most remain susceptible to the carbapenems (meropenem, imipenem, and ertapenem) [6].

AMINOGLYCOSIDES

Aminoglycosides are small hydrophilic and constitute one of the oldest class of antibiotics. Streptomycin was the first aminoglycoside discovered in 1914, and represents just one of a large class of antibiotics that share in common the general structure of two or more aminosugar joined by a glycosidic linkage to either hexose or aminoglycitol. Aminoglycosides are polycationic inhibitors of the 30S ribosomal subunit that interferes with protein synthesis and disrupts cell membrane transport and cell permeability. These drugs are most commonly used for the treatment of life-threatening systemic infections with Gram-negative bacilli, such as *Escherichia coli, K. pneumoniae, P. mirabilis*, and *P. aeruginosa* as well as used synergistically with ampicillin for *Enterococcus*.

While many aminoglycoside antibiotics have been developed and marketed worldwide over the years, the ones most commonly used in the United States today are amikacin, gentamicin, and tobramycin. Due to their known toxicity, which is similar to vancomycin (i.e., nephrotoxicity and ototoxicity), the use of this class of antibiotics has been largely reserved for treating the most serious infections, which may explain the low incidence of drug resistance seen thus far in comparison to many other more commonly used antibiotics. Other aminoglycosides, such as streptomycin and kanamycin, have been shown to be effective in the treatment of mycobacterial infections.

Aminoglycosides are poorly absorbed from the gastrointestinal tract. Therefore, preferred method of administration is intravenous infusion. All drugs in this class demonstrate similar volumes of distribution and overall kinetics with elimination half-life of 2–3 hours. These drugs exhibit little protein binding (<10%) and distribution can be increased, with corresponding decrease in peak levels measured in serum or plasma, in patients with fever and sepsis.

The major route of elimination for aminoglycosides is through the kidney, and these drugs remain essentially (85%–95%) unchanged because they are not metabolized by cytochrome P-450 mixed function oxidase, the major family of drug metabolizing enzymes mostly present in the liver. Patients with impaired renal function and elderly patients eliminate aminoglycosides more slowly and have longer drug half-lives when compared to normal adult population. In contrast, children and patients with fever have shorter half-life and lower concentrations of some aminoglycosides, particularly gentamicin. Patients with other conditions, such as cystic fibrosis, may also have lower serum or plasma concentrations due to increased renal clearance combined with larger volumes of distribution.

Aminoglycosides exhibit concentration-dependent bacteriocidal activity. Two predictors seem to be most closely correlated with efficacy: The area under the concentration curve (AUC) at 24 hours to minimum inhibitory concentration (MIC) ratio (AUC_{24}/MIC) and the ratio of peak aminoglycoside concentration (Cmax) to MIC. The Cmax to MIC ratio of 8–10 should be targeted for successful therapy with aminoglycosides and the Cmax can be established from known MIC data. Because clearance of aminoglycosides is proportional to glomerular filtration rate, patients with renal impairment may be more susceptible to aminoglycoside toxicity due to reduced clearance of aminoglycosides. For these patients, expansion of dosing frequency has been recommended. On the other hand, shorter dosing frequency may be needed for patients with higher clearance of aminoglycosides, for example, burn patients. The goal of dosing is to achieve maximum antimicrobial effect from these drugs.

Therapeutic Drug Monitoring of Aminoglycosides

Therapeutic monitoring of aminoglycosides is dependent on the type of dosing scenario that is used, traditional dosing, once daily (extended) dosing and synergy dosing. The primary purpose of monitoring in each case is to attain the desired therapeutic concentrations and to avoid toxic serum concentrations. However, if aminoglycoside therapy is conducted for three days or less, therapeutic drug monitoring is not required for most patients.

Peak concentrations are measured in serum or plasma specimens collected approximately 30 minutes after completion of infusion, whereas trough concentrations are collected approximately 15–30 minutes prior to next dose. Steady-state levels are reached after four or five half-lives. Once steady state and desired concentrations are attained, monitoring trough levels should performed at least weekly unless renal function is unstable or other nephrotoxic drugs are included. Typically, peak and trough concentrations 5–10 μg/mL and <2 μg/mL, respectively, for both gentamicin and tobramycin, whereas peak levels of 20–35 μg/mL and trough levels of 4–8 μg/mL are desired for amikacin in order to treat life-threatening infection. These targets can vary (increase or decrease) depending on type of infection being treated. Lower peak (15–25 μg/mL) and trough concentration (1–4 μg/mL) may be desirable in treating less severe infection.

For once daily dosage of aminoglycosides, monitoring peak concentration is redundant because desired peak concentration can be easily achieved with once daily dosing. However, for a patient where the volume of distribution could be significantly different than normally expected, for example, critically ill patients, burn patients, etc., measuring peak concentration may have some merit even with once daily dosing protocol.

Moreover, following single dose protocol of aminoglycosides, trough concentration could be below the detection limit of the assay and trough concentration determination may not be useful. Therefore, for once daily dosage of aminoglycosides there is no universally accepted guideline for monitoring. For 36–48 hours dosing interval in renally compromised patients, trough concentration monitoring has been suggested to ensure administration of another dosage does not cause aminoglycoside toxicity. Several nomograms have been published to aid dosing of aminoglycosides. These nomograms include MacGowan and Reeves, Begg and Nicolau nomograms. Another approach is to use Bayesian adaptive feedback software by a skilled pharmacologist to facilitate achievement of Cmax/MIC ratio of 8–10 and AUC targets of 70–120 mg h/L over 24 hours [7].

The use of aminoglycosides synergistically with other antibiotics is a common practice in treating Gram-negative and Gram-positive bacterial infections. For Gram-negative infections, monitoring should occur as described for traditional dosing approach. With aminoglycoside dosing for Gram-positive infections, lower doses will be required and minimal monitoring is recommended. In these instances, only trough levels should be monitored as peak concentrations should not be excessive—typically, for a drug like gentamicin, this should be maintained at 0.5–1 μg/mL. With aminoglycoside therapy, other tests monitored include serum creatinine should be measured at least twice weekly, with vestibular function/hearing tests also being indicated during prolonged courses of therapy (7–14 days). Suggested therapeutic range of vancomycin, aminoglycosides and selected antibiotics are given in Table 7.3.

Sustained peak concentrations of 15–20 μg/mL are associated with nephrotoxicity, while levels >16 μg/mL are associated with ototoxicity. Nephrotoxicity risks have been reported to increase in some patients with chronic liver disease and hypoalbuminemia. A less commonly encountered toxicity is neuromuscular blockade and aminoglycoside levels should be closely monitored when being coadministered with neuromuscular blocking agents.

VANCOMYCIN

Vancomycin is a tricyclic glycopeptide isolated from *Streptomyces orientalis*, which has time-dependent bacteriocidal activity against susceptible Gram-positive organisms. Vancomycin is a relatively large molecule with a molecular weight of 1456 daltons. Vancomycin interferes with cell

Table 7.3 Pharmacokinetics and guidelines for therapeutic drug monitoring of aminoglycosides and vancomycin

Antibiotic	Half-life (h)	Protein binding	Clearance	Peak level (μg/mL)	Trough level (μg/mL)
Amikacin	2–3	<10%	Renal mostly unchanged drug	20–35	4–8
Gentamicin	2–3	<10%	Renal mostly unchanged drug	5–10	<2
Tobramycin	2–3	<10%	Renal mostly unchanged drug	5–10	<2
Vancomycin	0.5–1	Approximately 55%	Renal mostly unchanged drug	20–40	10–15

well biosynthesis and is widely used to treat infections caused by viridans streptococci, coagulase negative staphylococci, *Enterococcus* species, and *S. aureus.* More recently, vancomycin has received considerable attention and has become the primary drug of choice against the growing problem of community and hospital-acquired MRSA as well as US 300, a community associated strain of MRSA. With greater and more widespread clinical use of vancomycin, the emergence of vancomycin-resistant enterococcal infections has risen dramatically, and is increasingly recognized as a major infection control problem amongst both hospitalized and institutionalized (long-term care) patient populations.

Oral vancomycin is poorly absorbed but oral vancomycin therapy is used for treating *Clostridium difficile* infection. Most common route of administration of vancomycin is intravenous administration. Therapeutic drug monitoring of vancomycin has been recommended for decades primarily to assess both therapeutic benefit and toxicity, most notably, ototoxicity (long-term vestibular and sensory damage) and nephrotoxicity. Vancomycin is protein bound in serum or plasma primarily to albumin in a variable amount. The protein bound fraction is estimated to be approximately 55%. The half-life is 0.5–1 hours in most patients, with a volume of distribution of 0.39–0.9 L/kg; this, however, may be prolonged in renal failure patients. Additional factors that can affect vancomycin distribution and penetration into tissues and fluids include protein binding, tissue type, tissue perfusion, degree of tissue inflammation, and diabetes, among others.

Vancomycin is eliminated principally by glomerular filtration and drug clearance also closely correlates with creatinine clearance. Small amounts are also eliminated by renal tubular secretion and through the biliary tract. Drug half-life increases as glomerular filtration declines. Certain age groups (neonates and children) and patient populations (burn victims and obese adults) have been observed to exhibit differences in elimination and distribution of vancomycin, when compared to normal adults.

Vancomycin demonstrates time-dependent killing of bacteria. By keeping vancomycin concentrations above the MIC, optimization of bacterial killing at the site of the infection is achieved. Vancomycin typically displays concentration-independent activity (as best documented against *S. aureus*), with the AUC divided by the MIC as the primary predictive pharmacodynamic parameter for efficacy. Some pharmacodynamics studies have shown that maintaining area under the concentration curve above the MIC (AUC_{24}/MIC) of 400 or 125 are associated with a greater chance of treating *S. aureus* pneumonia or *Enterococcus faecium* infections, respectively. As such, when the AUC_{24}/MIC increases, the patient may need to be exposed to higher concentrations of vancomycin. To achieve this pharmacokinetic-pharmacodynamic target, it is now widely felt by consensus expert committees that larger vancomycin doses and high trough serum concentrations may be required [8].

Therapeutic Drug Monitoring of Vancomycin

The primary premise for monitoring and adjustment of serum vancomycin concentrations is based on the perceived need to achieve serum concentrations at some multiple above the MIC for the offending organisms while avoiding potential adverse effects. However, if vancomycin is administered for 5 days or less, no therapeutic drug monitoring may be needed. For almost four decades of clinical practice, monitoring vancomycin concentrations in serum or plasma consisted of measuring peak and trough levels during intermittent dosing. The desired range has been traditionally specified at 20–40 μg/mL for peak concentrations and 10–15 μg/mL for trough concentrations. Although increasing trough vancomycin concentration to 15–20 μg/mL has been proposed, this proposal has not been supported by clinical trials. Therefore, alternative therapy should be considered in patients with MRSA infections if the MIC of vancomycin is 2 mg/L or higher because increasing the dosage to achieve higher trough vancomycin concentration may also increase the potential of vancomycin toxicity.

Currently, risk of vancomycin therapy is considered as minimal if the dosage is 1 g (15 mg/kg) every 12 hours but higher rate of nephrotoxicity is encountered at a dosage of 4 g/day or higher. The relationship between serum concentrations and treatment success or failure in serious *S. aureus* infections has been clearly established (failure rates exceeding 60% for *S. aureus* displaying a vancomycin MIC value of 4 mg/L) and prompted recommendations to lower the breakpoint for susceptibility from 4 to 2 mg/L by the Clinical and Laboratory Standards Institute and the Food and Drug Administration in 2006 and 2008, respectively. A number of studies have established a relationship between vancomycin treatment failures and infections in patients with MRSA displaying an MIC at 2 mg/L (2 μg/mL) or more. Taking into account other variables, for most Gram-positive organisms that have MIC values of 1 μg/mL, the desired trough concentration should be at least 8–10 μg/mL. For treatment of *S. aureus* pneumonia, additional recommendations have been made for trough concentrations of 10–15 μg/mL, with endocarditis, osteomyelitis, and meningitis requiring higher trough concentrations. Patel et al. studied pharmacodynamic profile of the more intensive dosing of vancomycin ranging from 0.5 mg intravenous every 12 hours to 2 g every 12 hours to treat MRSA infections using Monte Carlo stimulation. At an MIC of 2 mg/L, even the most aggressive vancomycin dosing regimen considered (2 g every 12 hours) only yielded a probability of target attainment (PTA) of 57% while generating a nephrotoxicity probability above 35%. But at MIC less than 1 mg/L, all subjected with trough vancomycin concentrations between 15 and 20 μg/mL, showed PTA of 100%. The authors concluded that vancomycin may not be useful for treating serious MRSA infections with MIC values exceeding 1 mg/L. Because AUC/MIC ratio of 400 or higher is the target associated with efficacy, incorporating AUC (area under the serum concentration curve) in routine therapeutic drug monitoring of vancomycin could be useful [9]. However, for all practical purposes, trough serum vancomycin concentrations are now viewed as the most accurate and practical method of monitoring the effectiveness of vancomycin therapy. Trough serum concentrations should be obtained just before the fourth dose, at steady-state conditions.

Mechanism of Resistance of Aminoglycosides and Vancomycin

There are three mechanisms of aminoglycoside resistance:

- Decreased uptake or altered permeability of aminoglycosides through bacterial cell membrane.
- Altered ribosomal binding site.
- Enzymatic modification of aminoglycosides (most common mechanism).

The bacterial cell wall serves as a natural barrier for small molecules such as aminoglycosides and altered permeability of the bacterial cell wall due to genetic mutations in the resistant bacteria. Moreover, efflux pump (P-glycoprotein mediated efflux mechanism) present on the bacterial cell wall may expel aminoglycosides trying to diffuse through the bacterial cell wall. Modification of efflux pump may further enhance bacterial resistance to aminoglycosides. Mutations in the ribosomal target of aminoglycoside, although rare may also explain bacterial resistance to antibiotics. However, resistance caused by bacterial 16S ribosomal RNA methyltransferase is of clinical importance because this enzyme can modify aminoglycoside binding site of 16S ribosomal RNA due to methylation of a nucleotide present in the binding site. However, most common mechanism of resistance is inactivation of these drugs by three types of bacterial enzymes: acetyltransferase, phosphotransferase, and nucleotidyltransferase [10].

Vancomycin has been the most reliable antibiotic for treating MRSA but since 1996 resistance of *S. aureus* to vancomycin has been reported (first report in a Japanese patient). Since then VRSA has been reported in many countries including the United States. Typically the MIC for vancomycin against *S. aureus* is 0.5–2 μg/mL but in VRSA it could be as high as 16 μg/mL or more. The mechanism of drug resistance is thickening of the bacterial cell wall. The understanding of the mechanism of resistance is based on the study of vancomycin-resistant enterococci (VRE). Under normal condition of peptidoglycan synthesis in enterococci, two molecules of D-alanine are joined by a ligase enzyme to form D-alanine–D-alanine which is the building block of peptidoglycan layer. Vancomycin binds with high affinity to D-alanine–D-alanine terminus thus preventing crosslinking. However, in resistant bacteria, due to genetic mutation (vanA gene) the precursor becomes D-alanine–D-lactate instead of D-alanine–D-alanine which has much lesser affinity of binding with vancomycin (class A resistance). The class B resistance is due to vanB gene which also produces D-alanine–D-lactate ligand. However, lower level of resistance is due to vanC (class C resistance) is due to D-alanine–D-serine ligand. Most

VRSA-positive patients have a history of infections caused by VRE containing vanA gene and MRSA. It is assumed that such genes may be transferred via plasmids or transposons from VRE to the MRSA strain resulting in VRSA [11].

DAPTOMYCIN AND LINEZOLID

Daptomycin is a novel cyclic lipopeptide antibiotic derived from *Streptomyces roseosporus* as a fermentation product. The drug must be administered intravenously once a day and was approved by Food and Drug Administration (FDA) in 2003. Daptomycin has bactericidal activity against Gram-positive bacteria including methicillin-susceptible *S. aureus*, MRSA, VRSA, penicillin-resistant *S. pneumoniae*, and ampicillin as well as VRE. This drug is also effective in treating complicated skin and skin-structure infections. Daptomycin binds to the bacterial cell membrane via calcium-dependent insertion of its lipid tail, forming an ion-conduction structure which rapidly depolarizes the cell membrane via efflux of potassium and possibly other ions. As a result there is a disruption of DNA, RNA, and protein synthesis which eventually causes bacterial cell death. The standard dose is 4 mg/kg in a 24 hours period. Dose should be lower in renally compromised patients (creatinine clearance <30). The drug has a good safety profile and low potential for resistance [12]. The drug is not approved for use in treating pneumonia due to its interaction with pulmonary surfactant, which results in inhibition of the antibiotic.

Linezolid is the first member of a new class of antibiotic drugs, the oxazolidinones. It has a broad range of activities against various Gram-positive bacteria including MRSA, VRE, and penicillin-resistant *S. pneumoniae*, but is bacteriostatic against these organisms. Linezolid also has some activity against atypical bacteria including some Nocardia and rapidly growing mycobacteria. Linezolid can be administered orally 400 mg or 600 mg twice daily. Linezolid is well tolerated with good safety record. Thrombocytopenia was documented as a side effect in only 2% of patients [13].

SULFONAMIDE AND TRIMETHOPRIM

Sulfonamides (sometimes called simply sulfa drugs), along with penicillin, are among the oldest and, historically, most widely prescribed chemotherapeutic agents for the treatment of infectious diseases. This large and diverse group of antimicrobial agents (which include sulfamethoxazole, sulfisoxazole, sulfadiazine, and many others) act as competitive inhibitors of dihydropteroate synthetase, a key enzymatic step in folate synthesis, to exhibit a bacteriostatic rather than bactericidal effect by interfering with normal cell division. These drugs are not used alone clinically except for sulfadiazine.

Trimethoprim is another bacteriostatic antibiotic, mainly used in the prophylaxis and treatment of urinary tract infections that acts by interfering with the action of bacterial dihydrofolate reductase. While used as monotherapy, trimethoprim has been commonly used in prescription combination (Bactrim, Septra) with the sulfonamide antibiotic, sulfamethoxazole. This co-trimoxazole therapy results in a synergistic antibacterial effect by inhibiting successive steps in the pathway of folate synthesis. Sulfamethoxazole-trimethoprim preparations are effective for treatment of infections due to chlamydia, susceptible strains of *Enterobacteriaceae* species such as *E. coli*, *Klebsiella* species, *Morganella morganii*, *P. mirabilis*, and *Proteus vulgaris*, in addition to Gram-positive cocci such as *S. aureus*, *Streptococcus pyogenes*, *S. pneumoniae*, and *Viridans streptococci*.

QUINOLONES

Quinolones, also known as fluoroquinolones, are a group of bactericidal antibiotics that are increasingly prescribed among inpatient and outpatient populations, largely due to their broad-spectrum actions against both Gram-positive and Gram-negative bacteria. These agents function by inhibiting DNA gyrase, a type II topoisomerase, and topoisomerase, which is an enzyme necessary to separate replicated DNA and, thereby, interferes with cell division. Drugs in this class include levofloxacin and ciprofloxacin, and may be administered via intravenously and oral routes.

Levofloxacin and ciprofloxacin are effective in treating infections (particularly tenacious and recurring respiratory tract infections) caused by Gram-positive pathogens (*S. pneumoniae* and *S. pyogenes* and some activity against *S. aureus* and *E. faecalis*), Gram-negative pathogens (*E. coli*, *H. influenzae*, *Moraxella catarrhalis*, *K. pneumoniae*, and *P. aeruginosa*), and atypical pathogens (*Chlamydia pneumoniae*, *Mycoplasma pneumoniae*, and *Legionella pneumophila*). Ciprofloxacin has also been increasingly used for treatment of both complicated and uncomplicated urinary tract infections, as resistance has emerged to many primary therapeutic agents such as sulfonamides and penicillins.

Toxicities may include diarrhea, gastrointestinal upset, allergic reactions, and phototoxicity. Central nervous system (CNS) effects such as agitation, restlessness, anxiety, and nightmares can also occur. QT prolongation has also been observed and rare cases of *torsades de pointes*, an uncommon type of ventricular tachycardia, have been reported. Achilles tendon rupture due to fluoroquinolone use is typically associated with renal failure.

MACROLIDES AND LINCOSAMIDES

The macrolides are a unique class of antibiotics whose activity stems from the presence of a macrolide ring, a large macrocyclic lactone ring to which one or more deoxysugar, usually cladinose and desosamine, may be attached. Erythromycin was one of the first drugs of this type used in clinical practice, but subsequent broader spectrum drugs, such as clarithromycin and azithromycin, have been developed more recently and are widely utilized. Like chloramphenicol, the site of action of these antimicrobial agents is inhibiting bacterial protein biosynthesis by binding reversibly to the subunit 50S of the bacterial ribosome and preventing translocation of peptidyl tRNA. Consequently, drugs in this class are primarily bacteriostatic, but can also be bacteriocidal when used in high concentrations.

The antimicrobial spectrum of macrolides is broader than that of penicillins (and comparable to many late generation cephalosporins); therefore, macrolide antibiotics have been successfully used as a substitute in treating patients with known penicillin allergy. Macrolide antibiotics are most frequently used to treat infections of the upper and lower respiratory tract and skin and soft tissue infections. They have been found to be particularly effective against beta-hemolytic streptococci, pneumococci, and staphylococci, in addition to many atypical respiratory pathogens such as *M. pneumoniae* and *L. pneumophila.* Nonetheless, they have also been used to treat chlamydia, syphilis, acne, gonorrhea, and nongonococcal cystitis, as well as certain mycobacterial and rickettsial infections.

Most common side effects are minimal and nonlife threatening, including gastrointestinal symptoms (diarrhea, nausea, abdominal pain, and vomiting) and headaches, dizziness, nervousness, rashes, and occasional facial swelling. Adverse reactions with macrolide antibiotics are relatively rare, and most do not require discontinuance of drug or routine therapeutic monitoring of drug concentrations in serum or plasma. However, serious allergic and dermatologic reactions have been reported with some drugs in this class, with the most serious and potentially life-threatening cases being toxic epidermal necrolysis and Stevens–Johnson syndrome (clarithromycin).

Mechanism of Resistance and D-Test

Macrolide antibiotics and lincosamide antibiotics differ structurally but share a similar mechanism of action. Resistances to these antibiotics are due to three different mechanisms:

- Modification of binding site by methylation or mutation that prevents the binding of these drugs with its ribosomal target in bacteria.
- Expelling antibiotic from bacterial cells by efflux mechanism.
- Producing specific enzymes that deactivate these drugs.

Modification of the ribosomal targets confers broad-spectrum resistance to macrolides and lincosamides whereas efflux mechanism and drug inactivation may only affect some specific antibiotics in these groups. Biochemical studies have shown that bacterial resistance to macrolide, lincosamides, and streptogramin B, the so-called MLS_B phenotype, is due to methylation of the ribosomal target of antibiotics. So far ribosomal methylation remains the most widespread mechanism of resistance to macrolides and lincosamides. Four major classes of genes responsible for such resistance are *erm*(A), *erm*(B), *erm*(C), and *erm*(F). The *erm*(A) and *erm*(C) typically are staphylococcal gene class, *erm*(B) class genes are found in streptococci and enterococci while *erm*(F) class genes are commonly found in *Bacteroides* species and other anaerobic bacteria [14].

D-test is a simple test to determine bacterial resistance to MLS_B. This test can be performed by disk diffusion, broth microdilution, or Etest. Erythromycin (macrolide) and clindamycin (lincosamide) disks can be placed adjacent to each other on the Mueller–Hinton agar medium that has been inoculated with the test organism. The objective is to look for blunting of the zone of inhibition around the clindamycin disk or strip. The normal zone around disks should be round, but in the presence of erythromycin-induction, the zone becomes D-shaped (flattening close to the clindamycin disk or strip). This indicates positive D-test meaning the bacteria is resistant to clindamycin and such resistance is indued by erythromycin (acquired resistance). If no flattening occurs (D-test negative) then bacteria is not resistant to clindamycin. This can also be achieved by combining the drugs in broth microdilution trays and looking for inhibition of the drug in broth.

TETRACYCLINES

Tetracyclines have been used since 1950s for treating a variety of Gram-positive and Gram-negative infections. In addition, tetracyclines are also used to treat infections due to intracellular chlamydiae, mycoplasmas, rickettsiae, and protozoa parasites. Tetracycline antibiotics are inexpensive making them attractive for use in developing countries. Tetracyclines can also be used prophylactically and also for the treatment of community-acquired infections especially respiratory infections. Ten different tetracycline derivatives have been marketed but doxycycline and minocycline are more commonly prescribed and can be administered orally. Tigecycline was developed later to overcome tetracycline resistance. Tigecycline is administered intravenously.

Tetracyclines reversibly inhibit bacterial protein synthesis by binding to the ribosomal complex thus preventing the association of aminoacyl-tRNA with bacterial ribosome. The cause of tetracycline resistance is due to acquisition of tetracycline resistance genes. More than 33 genes have been characterized and 23 of these genes encode efflux pump which are responsible for expelling antibiotics from bacterial cells. Ten other genes code for ribosomal protection proteins with very low affinity to bind with tetracycline drugs. These genes mostly confer resistance to tetracycline, doxycycline, and minocycline [15].

Tigecycline overcomes the two major resistance mechanisms of tetracycline; drug specific efflux pump acquisition by the bacteria and ribosomal protection. Tigecycline is effective against many Gram-positive and Gram-negative organisms including MRSA, vancomycin intermediate and VRE, and ESBL producing *E. coli* and *K. pneumoniae*. It is also active against many anaerobic bacteria as well as atypical pathogens including rapidly growing nontuberculous mycobacteria. Tigecycline is eliminated primarily through biliary excretion. Therefore, impaired renal function has no effect on its clearance. Common side effects are nausea and vomiting [16].

CARBAPENEMS

Carbapenems are beta-lactam antibiotics which have a broad spectrum of activity against many Gram-positive and Gram-negative aerobic and anaerobic organisms and are used for treating life-threatening serious infections not responding to standard antibiotic therapy. Carbapenems bind PBPs disrupting the integrity of the bacterial cell wall thus causing cell death. The mechanism of action is similar to beta-lactam antibiotics and cephalosporins but carbapenems have a fused beta-lactam ring structure which is resistant to most beta-lactamases produced by bacteria to inactivate penicillin like antibiotics. Carbapenems are very effective in treating infections caused by streptococci, enterococci, staphylococci, *Listeria*, Enterobacteriaceae, and many *Pseudomonas, Bacteroides*, and *Acinetobacter* species. MRSA are resistant to carbapenems. Carbapenems are also very effective against cephalosporin-resistant bacteria which are ESBL producers.

Imipenem was the first carbapenem approved for clinical use but this drug must be used in combination with cilastatin because imipenem is hydrolyzed in the human kidney by a dehydropeptidase enzyme to a nephrotoxic metabolite but cilastatin a dehydropeptidase inhibitor prevents formation of such nephrotoxin metabolite. Meropenem is however stable to human dehydropeptidases and does not require co-formulation with cilastatin. Other carbapenems are doripenem and ertapenem which can also be administered alone. All carbapenems must be administered intramuscularly or intravenously due to poor absorption from the gastrointestinal tract. As with other beta-lactam antibiotics, the most important adverse reaction is drug hypersensitivity. Mild elevation of liver enzymes may also be observed with carbapenem therapy.

Mechanisms of Resistance

Carbapenem-resistant Enterobacteriaceae (CRE) are a family of bacteria which are very difficult to treat because they are highly resistant to antibiotics. In general, common bacteria such as *Klebsiella* species and *E. coli* can become carbapenem resistant. This is due to bacterial synthesis of a carbapenemase enzyme that deactivates carbapenems by breaking down the beta-lactam part of the molecules. In the United States, the major challenge is presented by organisms that produce a *K. pneumoniae* carbapenemase (KPC) enzyme causing carbapenem resistance, but other types carbapenemase enzyme such as New-Delhi metallo-beta-lactamase and Verona-Integron-mediated metallo-beta-lactamase have also been reported in carbapenem-resistant bacteria [17]. Healthy people usually do not get CRE infection. These infections occur in hospitals, nursing homes, and other health care settings. Moreover, patients receiving long courses of antibiotics are susceptible to CRE infection which may be life-threatening due to limited treatment options. Certain nosocomial pathogens such as *Stenotrophomonas maltophilia* are inherently resistant to carbapenems due to the production of a metallo beta-lactamase.

COLISTIN/POLYMYXIN B

Polymyxins are a group of cationic polypeptide antibiotics consisting of five different compounds (polymyxin A–E). Colistin, also known as polymyxin E, was the first drug in this class which was discovered in 1949.

Currently, only colistin and polymyxin B are used clinically and both drugs are produced by *Bacillus* spp. The intravenous formulations of colistin and polymyxin B were used during 1960s but these drugs were getting less popular due to their toxicities. However, in recent years, these drugs were reintroduced due to the emergence of multidrug resistance in Gram-negative bacteria. The polymyxins are active against Gram-negative bacteria such as *Acinetobacter baumannii*, *P. aeruginosa*, and *K. pneumonia* strains which are multidrug-resistant. These bacteria are responsible for intensive care unit-acquired infections such as pneumonia/ventilator-associated pneumonia, central venous catheter-related infection, urinary tract infection, meningitis, and bacteremia/sepsis. The polymyxins exert their bactericidal activity by binding to the bacterial cell membrane and disrupting its permeability that results in leakage of intracellular components. Dosage of colistin in critical care patients is usually 2.5–5.0 mg/kg/day (given in a divided dosage of 2–4 times in a day) while dosage of polymyxin B is 1.5–2.5 mg/kg/day, administered in divided dosage. The major adverse effects are nephrotoxicity, neurotoxicity, and neuromuscular blockage [18].

ANTIMYCOBACTERIAL AGENTS

Antimycobacterial, or antituberculosis, agents represent a diverse group of structurally unrelated compounds that include rifampin, isoniazid, ethambutol, streptomycin, amikacin, kanamycin, and pyrazinamide. Other antibiotics may be added to the regimen if needed, for example, a fluoroquinolone. Each of these agents has very different and unique biological mechanisms of action; yet, when used alone or in combination, these drugs have historically proven to be clinically effective in the acute treatment and long-term public health management of *Mycobacterium* infections such as tuberculosis and leprosy. Antimycobacterial agents are most commonly prescribed today in multidrug combinations due to the frequent emergence of resistance, which is often seen as a consequence of poor patient compliance and incomplete treatment of active infections during the long course required with most therapeutic regimens.

Rifampin inhibits DNA-dependent RNA polymerase in bacterial cells by binding to its beta-subunit and, thus, preventing RNA transcription and subsequent translation to proteins. In contrast, isoniazid and ethambutol separately exert their bacteriostatic effects through different molecular mechanisms by interfering with the synthesis of the mycolic acid–peptidoglycan complex in the cell wall to increase overall cell permeability.

Streptomycin, amikacin, and kanamycin are aminoglycoside antibiotics that interfere with protein synthesis and alter cell membrane transport, with a similar net effect of increasing overall cell permeability as well.

The most serious adverse and toxic side effects have been observed with rifampin and isoniazid, most notably, hepatitis and jaundice (with liver failure in severe cases) and sideroblastic anemia. Other milder side effects include flushing, pruritus, rash, redness, and watering of eyes, as well as gastrointestinal and CNS disturbances and general flu-like symptoms. As observed with most other bacteriostatic agents, toxicity with these drugs is independent of drug concentration. Serious drug toxicity was reported in one 37-year-old woman who received ethambutol 825 mg, isoniazid 225 mg, rifampicin 450 mg, and pyrazinamide 1200 mg and had serious impairment in her vision. The authors commented that ethambutol and isoniazid are notorious for causing visual impairment [19].

Pyrazinamide is an analog of nicotinamide which is used as an antituberculous agent (bacteriostatic or bactericidal) most often used in combination with other antituberculosis agents for the initial treatment of active tuberculosis in children and adults. Pyrazinamide (500 mg tablet) is administered orally because it is well absorbed from the gastrointestinal tract. The half-life is 9–10 hours and approximately 70% of this drug is excreted in urine unchanged within 24 hours. Pyrazinamide inhibits excretion of uric acid by the kidney and may elevate serum uric acid concentration. This may not be a problem of most patients but pyrazinamide should be used with caution in patients with gout. Therapeutic drug monitoring of pyrazinamide is not recommended.

Although antimycobacterial agents are generally not subjected to therapeutic drug monitoring, treatment failure may result from sub-therapeutic drug concentrations. Peloquin commented that therapeutic drug monitoring may not be needed for otherwise healthy individuals who are responding to standard four drug treatment regimens for tuberculosis, but therapeutic drug monitoring is useful in patients who are slow to respond to therapy or at a higher risk of drug–drug interactions. Early intervention based on therapeutic drug monitoring is useful for these patients to avoid development of drug resistance. When only one specimen is obtained for therapeutic drug monitoring 2 hours postdosage specimen is appropriate for isoniazid, rifampin, ethambutol, and pyrazinamide [20]. Babalik et al. reported that 60% of patients receiving antituberculosis drugs in their study showed below therapeutic drug concentration. The authors commented that drug levels were below therapeutic ranges in patients with active tuberculosis particularly in patients with HIV infection or other comorbidities [21].

OTHER ANTIBIOTICS

Metronidazole is a rare example of a drug developed against a parasite that gained broad use as an antibiotic. The antibacterial activity of metronidazole was discovered by accident in 1962 when metronidazole cured a patient suffering from both trichomonad vaginitis and bacterial gingivitis. However, it was not till 1970s when metronidazole gained widespread acceptance from clinicians for treating infections caused by Gram-negative anaerobes such as Bacteroides or Gram-positive anaerobes such as Clostridium. Metronidazole is active against *Entamoeba histolytica* the cause of amebic dysentery and *Giardia lamblia.* Currently metronidazole is used as a prophylaxis against anaerobic infection after bowel surgery, and also an important part of combination therapy against *Helicobacter pylori*, a major cause of gastritis and a risk factor for stomach cancer [22].

In 2014, FDA approved Zerbaxa (ceftolozane/tazobactam), a novel combination of an antipseudomonal cephalosporin and a beta-lactamase inhibitor. Ceftolozane inhibits formation of bacterial cell wall by binding with PBP while tazobactam irreversibly inhibits the activity of bacterial enzymes (beta-lactamase and cephalosporinase) that may inactivate ceftolozane. The FDA approved this drug for treating adults with complicated intra-abdominal infections and complicated urinary tract infections including pyelonephritis due to multidrug-resistant (MDR) *P. aeruginosa.* In 2015, FDA approved another new antibiotic agent avycaz (ceftazidime/avibactam) for treatment of complicated intra-abdominal infection in combination with metronidazole and complicated urinary tract infection including kidney infection pyelonephritis. Ceftazidime is a previously approved cephalosporin drug which in this combination is added to avibactam, a new beta-lactamase inhibitor. This drug has activity against MDR Enterobacteriaceae that are resistant to carbapenems.

ANTIFUNGAL AGENTS

The population of patients at risk of invasive mycoses has been increasing as a result of infection, malignancies, as well as organ transplantation because these patients are on lifelong immunosuppressant therapy. In addition, patients with severe HIV infection are immunocompromised.

Candida and *Aspergillus* species remain the major cause of fungal infection. Approximately 50% of patients with candidiasis are infected with *Candida albicans* while *Aspergillus fumigatus* is responsible for most cases of invasive aspergillosis. The oldest drug for treating fungal infection is amphotericin B which belongs to the polyene class. Other drugs in this class are nystatin and natamycin. These drugs are produced by Streptomyces bacteria. Polyenes bind to ergosterol in fungal plasma membrane causing increased permeability and fungal cell death. Amphotericin B has antifungal activity against a wide variety of yeasts and molds including *Candida* and most *Aspergillus* species. Conventional amphotericin B administration is associated with nephrotoxicity and infusion-related reactions just limiting its use. Lipid formulation of amphotericin B such as amphotericin B lipid complex, liposomal amphotericin B, and amphotericin B colloidal complexes were developed to reduce nephrotoxicity and infusion-related reactions. These are the formulation available currently for clinical use.

The azole group of antifungal agents: fluconazole, itraconazole, and newer triazoles such as voriconazole and posaconazole are widely used today for treating various fungal infections. Azoles inhibit synthesis of fungal ergosterol by inhibiting enzyme lanosterol 14-alpha-demethylase. As a result fungal cell membranes are disrupted causing fungal death. Azoles are effective in treating fungal infection due to various *Candida* and *Aspergillus* species. These drugs can be administered orally but these drugs are metabolized by liver cytochrome P-450 mixed function enzymes and as a result may pharmacokinetically interact with many other drugs. Therapeutic drug monitoring is useful in monitoring therapy with azoles as well as potential drug–drug interactions. Fluconazole and voriconazole have high oral bioavailability and they also attain high concentration in cerebrospinal fluid. Hepatotoxicity may occur but posaconazole has the least hepatotoxicity.

Echinocandins (caspofungin, micafungin, and anidulafungin) are more recently introduced antifungal drugs which inhibit fungal cell wall formation by inhibiting 1,3-beta-D-glycan synthase, the enzyme responsible for the production of 1,3-beta-D-glycan, an essential component of the fungal cell wall. These drugs are fungicidal against most yeast species and fungistatic against molds. These drugs have poor bioavailability and must be administered intravenously. These drugs have less toxicity than azoles and minimal interactions with liver cytochrome P-450 mixed function enzymes. As a result drug–drug interactions are minimal and therapeutic drug monitoring is not necessary [23].

Therapeutic Drug Monitoring of Azoles

Fluconazole is well absorbed after oral administration and pharmacokinetics is linear. Therefore, routine therapeutic drug monitoring may not be necessary. However, The British Society of Medical Mycology recommends that a trough serum itraconazole concentration should be determined 5–7 days after initiation of therapy of dosage adjustment and a target serum level of 0.5–1.0 mg/L itraconazole is recommended. Serum voriconazole levels correlate with both clinical efficacy and toxicity and therapeutic drug monitoring of voriconazole is also recommended. A target trough concentration of >1 mg/L is recommended for both prophylaxis and therapy. A higher target of 2 mg/L may be used treating severe fungal infection with poor prognosis such as infection with CNS or multifocal infection. Usually toxicity is observed at serum concentration above 5.5 mg/L. For posaconazole, the target trough concentration is >0.7 mg/L measured 7 days after initiation of therapy [24]. Chromatographic methods such as high performance liquid chromatography combined with ultraviolet detection or liquid chromatography combined with tandem mass spectrometry are available for therapeutic drug monitoring of azoles.

ANTIBIOTIC SUSCEPTIBILITY TESTING

Antibiotic susceptibility testing (AST) is an in vitro measure to assess the likelihood that a particular antimicrobial agent will treat an infection caused by a particular organism. Cumulative AST data used to create an antibiogram can predict appropriate empiric antimicrobial therapy in a particular population prior to the availability of specific information on the patient's isolate. The basic methods of AST include a qualitative test or Kirby Bauer disk diffusion a quantitative method which can be in the form of broth macro or microdilution and can also be done in an automated instrument [25,26]. With either method interpretation of results is based on guidelines provided by the Clinical Laboratory Standards Institute (CLSI) [27]. In the Kirby Bauer method, paper disks with a specific concentration of antibiotic are placed on a lawn of bacteria. After incubation the diameter of the zone of inhibition is measured and results are reported as susceptible, intermediate, or resistant. This method of testing is often used in treatment of uncomplicated infections in normal hosts such as urinary tract infections. A quantitative method in which an MIC value and an interpretation of susceptible, intermediate, and resistant is reported is very important in treatment of serious infections especially in patients with compromised immune systems or those that are critically ill. The MIC represents the lowest concentration of an antimicrobial that inhibits the visible growth of a microorganism after overnight incubation. MICs are also important to confirm resistance of microorganisms to an antimicrobial agent and also to determine the potency of new antibiotics for which no CLSI interpretations are available.

Molecular methods can also be used to specifically determine the resistance mechanism of an organism. Molecular methods are the standard for testing viruses for susceptibility to antiviral agents as well (see Chapter 13: Application of Molecular Diagnostics).

PRINCIPLES OF ANTIBIOTIC SUSCEPTIBILITY TESTING

The concentration of organisms used in performing the susceptibility test is critical. An inoculum that is too heavy can result in false resistance and an inoculum that is too light can result in false susceptibility. A suspension is made from approximately five colonies of a pure culture of the organism to be tested growing on a nonselective media. The suspension is prepared in broth to a turbidity equivalent to a McFarland 0.5, which is $1–2 \times 10^8$ CFU/mL organisms. The inoculum must be diluted to a final concentration of 5×10^5 CFU/mL and plated on agar or into broth within 15 minutes of preparation.

The media used for susceptibility testing is Mueller–Hinton broth or agar. Agar plates must be those with a depth of 3–5 mm. If the media is too thin this could lead to false susceptibility and if the agar is too deep it could lead to zones that are too small (false resistance). Agar plates or broth is incubated at 35°C in ambient conditions for 16–20 hours. Mueller–Hinton can be supplemented with 5% sheep blood or lysed blood to support growth of fastidious organisms. Specific supplementation such as 2% NaCl for detection of oxacillin resistance in *Staphylococcus* species and incubation conditions such as 24 hours for oxacillin and vancomycin are required for accurate detection of specific resistance mechanisms.

Screening tests can also be used to detect specific resistance mechanisms. One such screening test is used to detect the presence of ESBLs in *E. coli, K. pneumoniae, Klebsiella oxytoca,* or *P. mirabilis*. Cefotaxime and ceftazidime with or without clavulanic acid is used in a disk or broth microdilution format. An increase in zone size by 5 or more mm with the one of the two combination drugs compared to the drug alone is indicative of the presence of an ESBL. Inducible resistance to clindamycin can be determined in *Staphylococcus* species, beta-hemolytic streptococci, or *S. pneumoniae* using the "D" test. An erythromycin disk or Etest strip is placed in close proximity to a clindamycin disk or strip. A flattening or blunting of the clindamycin zone on the side of the erythromycin is a marker for the presence of the *erm* gene in the strain and hence a positive D-test. Strains with a positive D-test should be considered resistant to clindamycin even if it appears susceptible by routine susceptibility testing methods. Detection of carbapenem-resistant *K. pneumoniae* can be performed by using the Hodge test. The principle of the Hodge test is that production of this carbapenemase in a strain of *Klebsiella* will allow growth of a susceptible strain of *E. coli* up close to an ertapenem disk in the form of a cloverleaf pattern.

Selection of drugs to be reported is dependent on the site of infection based on penetration of select antimicrobials into different sites of infection. Guidelines can be found in the newest CLSI documents. Some examples include drugs such as nitrofurantoin and fosfomycin should only be reported in the urine. Oral agents, clindamycin, macrolides, tetracycline, and fluoroquinolones should not be reported in CSF samples due to the lack of utility in the treatment of meningitis.

Antimycobacterial Susceptibility Testing

Infections due to rapidly growing mycobacteria, such as *Mycobacterium chelonae/abscessus* complex and *Mycobacterium fortuitum* complex are treated with antibacterial agents, but often in combination and for long periods of time. Although patterns of susceptibility can be species dependent the specific identification can be difficult and time consuming to perform, susceptibility testing is important for clinically significant organisms. Susceptibility can be performed as described for bacteria using either Etest or broth microdilution.

Susceptibility testing of *Mycobacterium tuberculosis* has historically been performed by agar proportion method or broth using a single concentration of antituberculous drugs. If the organism grows it is considered resistant and if it does not it is susceptible. No susceptibility is routinely performed on other slow growing mycobacteria. Susceptibility testing of *Mycobacterium avium-intracellulare* may be indicated in complicated patients, not responding to standard therapeutic regimens.

Susceptibility testing is important for the treatment of infections due to yeast including *Candida* species and *Cryptococcus neoformans*. The increase in the use of antifungal agents in hospitalized patients has led to an increase in the prevalence of Candida species nonalbicans which tend to be more resistant to azole antifungal agents as compared to *C. albicans*. Some generalizations include the fact that *Candida glabrata* is not often susceptible to fluconazole and if used for treatment should be a high dose, *Candida krusei* is always resistant to fluconazole. *Candida lusitaniae* can be resistant to amphotericin. Some *Candida parapsilosis* can have higher than normal MICs to echinocandin antifungal agents. Antifungal testing is recommended for clinically significant isolates. Testing for azole antifungal agents can be performed by disk diffusion, automated broth microdilution, or Etest although the media and conditions are different. Specific guidelines can be found in the CLSI document [27]. CLSI also has standards for susceptibility testing of filamentous fungi (molds) although no breakpoints are available, however reporting of MICs alone can be helpful in treatment of these infections which often occur in immunocompromised patients.

ANTIPARASITIC AGENTS

Malaria is a worldwide problem especially in tropical countries. Quinine isolated from bark of cinchona tree is the oldest drug for treating malaria. Chloroquine is the drug of choice as a prophylaxis and treatment of *Plasmodium* but chloroquine-resistant malaria is widespread in South East Asia and India. Amodiaquine is similar in structure of chloroquine but retains some activity against chloroquine-resistant malaria parasite. It is often used as in co-formulation with artesunate. Piperaquine is also useful in treating chloroquine-resistant malaria. Quinine has a short half-life and if given orally must be administered three times a day. Primaquine is also used in treating malaria but it may cause hemolytic anemia in people with glucose-6-phosphate dehydrogenase deficiency. Mefloquine is the drug of choice for use as a prophylaxis for travelers to endemic areas with chloroquine-resistant *Plasmodium falciparum*. This drug has a long half-life (15–33 days) and should be taken only once a week. Other antimalarial agents include dapsone (antifolate), doxycycline (used commonly as an antibiotic), and artemisinin (isolated from Chinese medicinal herb *Artemisia annua*).

There are several antiparasitic drugs. For example, benzimidazole class of antiparasitic agents includes thiabendazole, mebendazole, albendazole, and triclabendazole. Mebendazole is the drug of choice to treat most common nematode infections. Albendazole has activity against less common tissue nematode infections. Thiabendazole is an old drug but its side effects limited its use. Triclabendazole is the drug of choice for treating live fluke infection due to *Fasciola hepatica*. Ivermectin is a semi-synthetic derivative of avermectin. Ivermectin is the drug of choice for treating onchocerciasis and strongyloidosis. This drug is also active against other filarial worms. Praziquantel has activity against schistosomiasis, intestinal tapeworms, cysticercosis, and other flukes. Other agents for treating parasites include pyrantel, nitazoxanide, oxamniquine, metronidazole, diethylcarbamazine, and piperazine.

ANTIVIRAL AGENTS

Antiviral agents target virus and can be classified under two broad categories: Treating non-HIV viral infections and treating HIV. For treatment of influenza, four antiviral agents (amantadine, rimantadine, oseltamivir, and zanamivir) are available. Amantadine and rimantadine are active against influenza a virus while oseltamivir and zanamivir have activity against both influenza A and B viruses. For treating chronic hepatitis B infection, lamivudine, as well as adefovir, and dipivoxil are approved. For treating hepatitis C infection, combination of ribavirin with interferon-alpha is effective. Agents active against herpesvirus include acyclovir, valacyclovir, penciclovir (when applied topically), famciclovir, idoxuridine (applied topically), and trifluridine (applied topically). These agents are also useful in treating varicella zoster virus infections. Moreover, ganciclovir, valganciclovir, foscarnet, cidofovir, and formivirsen are effective in treating cytomegalovirus (CMV) in patients with AIDS [28].

Currently 28 drugs are approved for treating patients infected with HIV. Highly active antiretroviral therapy (HAART) provides effective treatment options for treatment-naive and treatment-experienced patients. Six classes of antiretroviral agents are currently available [29]:

- Nucleoside reverse transcriptase inhibitors: Zidovudine, didanosine, zalcitabine, stavudine, lamivudine, abacavir, tenofovir, and emtricitabine.
- Nonnucleoside reverse transcriptase inhibitors: Nevirapine, delavirdine, efavirenz, etravirine, and rilpivirine.
- Protease inhibitors: Saquinavir, indinavir, ritonavir, nelfinavir, amprenavir, lopinavir, fosamprenavir, atazanavir, tipranavir, and darunavir.
- Integrase inhibitors: Raltegravir, elvitegravir, and dolutegravir.
- Fusion inhibitors: Enfuvirtide.
- Chemokine receptor antagonists (CCR5 antagonists): Maraviroc.

Although conventional susceptibility testing of viruses can be performed by growing the virus in tissue culture and using antiviral agents to perform a plaque reduction assay resistance to these agents is most commonly performed by molecular methods such as ganciclovir resistance testing of CMV or antiretroviral testing (genotyping) for HIV. Specifics on the testing of these agents can be found.

KEY POINTS

- There are six major mechanisms by which an antibiotic exerts its pharmacological action. These included: Inhibition of cell wall synthesis, inhibition of bacterial protein synthesis, disrupting bacterial cell membrane, damaging bacterial cell membrane, inhibition of bacterial nucleic acid synthesis, and antimetabolite activities. Inhibition of bacterial cell wall formation is probably the most common mechanism by which an antibiotic kills bacteria or inhibits bacterial growth.
- The half-life of a drug is the time required for the serum concentration to be reduced by 50%. Therapeutic drug monitoring is recommended when a drug reaches a steady state which is approximately after five half-lives.
- Penicillins, such as Penicillin G, Penicillin V, amoxicillin, nafcillin, and ampicillin, are used for treatment of Gram-positive infections such as *Staphylococcus* species, *S. pneumoniae*, beta-hemolytic strains of *Streptococcus*, and *E. faecalis*. These drugs may be used alone or in preparations that include beta-lactamase inhibitors, such as clavulanic acid, tazobactam, and sulbactam, for treatment of penicillin-resistant organisms.
- Several generations (first through fifth) of cephalosporin antibiotics have now been developed and are widely prescribed, with an increasingly broader spectrum of action against both Gram-negative and Gram-positive organisms. Later generations of cephalosporins, such as cefaclor, ceftriaxone cefotaxime are effective against organisms including *P. mirabilis*, *Klebsiella* species, *E. aerogenes*, and *H. influenzae*, while ceftazidime and cefepime have activity against *P. aeruginosa*. Cefepime and ceftazidime are effective in treating Gram-negative nosocomial infections in critically ill patients.
- Beta-lactams and cephalosporins exert their antimicrobial effect by covalently binding with bacterial PBPs thus interrupting bacterial cell wall synthesis. Resistance to beta-lactam antibiotics may be due to alteration of PBPs with reduced no binding with antibiotic, decreased

ability of the antibiotic to diffuse through bacterial cell wall (porins), or bacteria may synthesize beta-lactamases enzymes that inactivate the antibiotic by hydrolysis of beta-lactam ring. ESBLs are a rapidly evolving group of beta-lactamases which are capable of hydrolyzing third generation cephalosporins and aztreonam but are inhibited by clavulanic acid. The ESBLs are frequently encoded by plasmid genes and these enzymes are responsible for extreme drug resistance in some bacteria.

- Aminoglycosides (amikacin, gentamicin, and tobramycin most commonly used) are used in treating life-threatening systemic infections with Gram-negative bacilli, such as *E. coli*, *K. pneumoniae*, *P. mirabilis* and *P. aeruginosa* as well as used synergistically with ampicillin for *Enterococcus*. These drugs are administered intravenously due to poor bioavailability. Therapeutic drug monitoring of amikacin and tobramycin is essential. Due to high toxicity, both peak and trough concentrations are measured. However, if aminoglycoside therapy is conducted for three days or less, therapeutic drug monitoring is not required for most patients.
- Major toxicities of aminoglycosides are ototoxicity (mostly irreversible) and nephrotoxicity (may be reversible).
- Vancomycin is a tricyclic glycopeptide isolated from *S. orientalis*, which has time-dependent bacteriocidal activity against susceptible Gram-positive organisms. Vancomycin is widely used to treat infections caused by viridans streptococci, coagulase negative *Staphylococcus*, *Enterococcus* species, and *S. aureus.* More recently, vancomycin has received considerable attention and has become the primary drug of choice against the growing problem of community and hospital-acquired MRSA. Due to high toxicity, therapeutic drug monitoring is essential where both peak and trough level should be measured. However, therapeutic drug monitoring may be avoided if the drug is used for five days or less.
- Major toxicities of vancomycin are ototoxicity and nephrotoxicity.
- Daptomycin is a novel cyclic lipopeptide antibiotic derived from *S. roseosporus* as a fermentation product. The drug must be administered intravenously once a day and was approved by FDA in 2003. Daptomycin has bactericidal activity against Gram-positive bacteria including methicillin-susceptible *S. aureus*, MRSA, VRSA, penicillin-resistant *S. pneumoniae*, and ampicillin as well as VRE. This drug is also effective in treating complicated skin and skin-structure infections.

- Linezolid is the first member of a new class of antibiotic drugs, the oxazolidinones. It has a broad range of activities against various Gram-positive bacteria including MRSA, VRE, and penicillin-resistant *S. pneumoniae*.
- Sulfonamides (sulfamethoxazole, sulfisoxazole, sulfadiazine, and many others), along with penicillin, are among the oldest and, historically, most widely prescribed chemotherapeutic agents for the treatment of infectious diseases. These drugs are not used alone clinically except for sulfadiazine.
- Trimethoprim is another bacteriostatic antibiotic, mainly used in the prophylaxis and treatment of urinary tract infections that acts by interfering with the action of bacterial dihydrofolate reductase. While used as a monotherapy, trimethoprim has been commonly used in prescription combination (Bactrim, Septra) with the sulfonamide antibiotic, sulfamethoxazole.
- Quinolones, also known as fluoroquinolones, are a group of bactericidal antibiotics that are increasingly prescribed among inpatient and outpatient populations, largely due to their broad-spectrum actions against both Gram-positive and Gram-negative bacteria.
- The macrolides are a unique class of antibiotics whose activity stems from the presence of a macrolide ring, a large macrocyclic lactone ring to which one or more deoxysugar, usually cladinose and desosamine, may be attached. Erythromycin was one of the first drugs of this type used in clinical practice, but subsequent broader spectrum drugs, such as clarithromycin and azithromycin, have been developed more recently and are widely utilized.
- D-test is very useful to identify inducible clindamycin resistance.
- Tigecycline was developed later to overcome tetracycline-resistant bacteria. Unlike many tetracyclines which can be administered orally, tigecycline is administered intravenously.
- Carbapenems are beta-lactam antibiotics which have a broad spectrum of activity against many Gram-positive and Gram-negative aerobic and anaerobic organisms and are used for treating life-threatening serious infections not responding to standard antibiotic therapy.
- Polymyxins are a group of cationic polypeptide antibiotics consisting of five different compounds (polymyxin A–E). Colistin, also known as polymyxin E, was the first drug in this class which was discovered in 1949. Currently, only colistin and polymyxin B are used clinically and both drugs are produced by *Bacillus* spp.

- Antimycobacterial, or antituberculosis, agents represent a diverse group of structurally unrelated compounds that include rifampin, isoniazid, ethambutol, streptomycin, amikacin, and kanamycin. Other antibiotics may be added to the drug therapy if needed, for example, a fluoroquinolone.
- *Candida* and *Aspergillus* species remain the major cause of fungal infection. Approximately 50% patients with candidiasis are infected with *C. albicans* while *A. fumigatus* is responsible for most cases of invasive aspergillosis. The oldest drug for treating fungal infection is amphotericin B which belongs to polyene class. Other drugs in this class are nystatin and natamycin. Azoles group of antifungal agents: Fluconazole, itraconazole, and newer triazoles such as voriconazole and posaconazole are widely used today for treating various fungal infections. Echinocandins (caspofungin, micafungin, and anidulafungin) are more recently introduced antifungal drugs which are fungicidal against most yeast species and fungistatic against molds. These drugs have poor bioavailability and must be administered intravenously.
- Therapeutic drug monitoring is useful for therapy with itraconazole, voriconazole, and posaconazole. Therapeutic drug monitoring of fluconazole is not usually performed unless for some very specific patients.
- AST is an in vitro measure to assess the likelihood that a particular antimicrobial agent will treat an infection caused by a particular organism. Cumulative AST data used to create an antibiogram that can predict appropriate empiric antimicrobial therapy in a particular population prior to availability of specific information on the patient's isolate. The basic methods of AST include a qualitative test or Kirby Bauer disk diffusion a quantitative method which can be in the form of broth macro or microdilution and can also be done in an automated instrument.
- The media used for susceptibility testing is Muellar–Hinton broth or agar. Agar plates or broth is incubated at 35°C in ambient conditions for 16–20 hours. Muellar–Hinton can be supplemented with 5% sheep blood or lysed blood to support growth of fastidious organisms. Specific supplementation such as 2% NaCl for detection of oxacillin resistance in *Staphylococcus* species and incubation conditions such as 24 hours for oxacillin and vancomycin are required for accurate detection of specific resistance mechanisms.
- Malaria is a worldwide problem especially in tropical countries. Chloroquine is the drug of choice as a prophylaxis and treatment of

Plasmodium but chloroquine-resistant malaria is widespread in South East Asia and India. Other antimalarial drugs include amodiaquine, piperaquine, primaquine (may cause hemolytic anemia in people with glucose-6-phosphate dehydrogenase deficiency), and mefloquine (half-life 15–33 days), dapsone (antifolate), doxycycline (used commonly as an antibiotic), and artemisinin (isolated from Chinese medicinal herb *A. annua*).

- There are several antiparasitic drugs. For example, benzimidazole class of antiparasitic agents includes thiabendazole, mebendazole, albendazole, and triclabendazole. Mebendazole is the drug of choice to treat most common nematode infections. Albendazole has activity against less common tissue nematode infections. Triclabendazole is the drug of choice for treating live fluke infection due to *F. hepatica*. Ivermectin is a semisynthetic derivative of avermectin. Ivermectin is the drug of choice for treating onchocerciasis and strongyloidosis. This drug is also active against other filarial worms. Praziquantel has activity against schistosomiasis, intestinal tapeworms, cysticercosis, and other flukes. Other agents for treating parasites include pyrantel, nitazoxanide, oxamniquine, metronidazole, diethylcarbamazine, and piperazine.
- For treating chronic hepatitis B infection, lamivudine, as well as adefovir, and dipivoxil are approved. For treating hepatitis C infection, combination of ribavirin with interferon-alpha is effective.
- Agents active against herpesvirus include acyclovir, valacyclovir, penciclovir (when applied topically), famciclovir, idoxuridine (applied topically), and trifluridine (applied topically).
- Currently 28 drugs are approved for treating patients infected with HIV. HAART provides effective treatment options for treatment-naive and treatment-experienced patients.

REFERENCES

[1] Spanu T, Santangelo R, Andreotti F, Cascio GL, et al. Antibiotic therapy for severe bacterial infections: correlation between the inhibitory quotient and outcome. Int J Antimicrob Agents 2004;23:120–8.

[2] Rhee KY, Gardiner DF. Clinical relevance of bacteriostatic versus bactericidal activity in the treatment of Gram-positive bacterial infections. Clin Infect Dis 2004;39:755–6.

[3] Roberts JA, Webb S, Lipman J. Cefepime versus ceftazidime: considerations for empirical use in critically ill patients. Int J Antimicrob Agents 2007;29:117–28.

[4] Poon H, Chang MH, Fung HB. Ceftaroline fosamil: a cephalosporin with activity against methicillin-resistance *Staphylococcus aureus*. Clin Ther 2012;34:743–65.

[5] Pitout JD, Sanders CC, Sanders WE Jr. Antimicrobial resistance with focus on beta-lactam resistance in Gram negative bacilli. Am J Med 1997;103:51–9.

[6] Paterson DL, Bonomo RA. Extended spectrum beta-lactamases: a clinical update. Clin Microbiol Rev 2006;18:657–86.
[7] Avent ML, Rogers BA, Cheng AC, Paterson DL. Current use of aminoglycosides: indication, pharmacokinetics and monitoring for toxicity. Int Med J 2011;41:441–9.
[8] Rybak MJ, Lomaestro BM, Rotscahfer JC, Moellering RC, et al. Vancomycin therapeutic guidelines: a summary of consensus recommendations from the Infectious Diseases Society of America, the American Society of Health-System Pharmacists, and the Society of Infectious Diseases Pharmacists. Clin Infect Dis 2009;49:325–7.
[9] Patel N, Pai MP, Rodvold KA, Lomaestro B, et al. Vancomycin: we can't get there from here. Clin Infect Dis 2011;52:969–74.
[10] Jana S, Deb JK. Molecular understanding of aminoglycoside action and resistance. Appl Microbiol Biotechnol 2006;70:140–50.
[11] Hiramatsu K. Vancomycin resistant *Staphylococcus aureus*: a new model for antibiotic resistance. Lancet Infect Dis 2001;3:147–55.
[12] Carpenter CF, Chambers HF. Daptomycin: another novel agent for treating infection due to drug-resistant gram positive pathogens. Clin Infect Dis 2004;38:994–1000.
[13] Perry CM, Jarvis B. Linezolid: a review of its use in the management of serious Gram positive infection. Drugs 2001;61:525–51.
[14] Leclercq R. Mechanism of resistance of macrolides and lincosamides: nature of resistance elements and their clinical implications. Clin Infect Dis 2002;34:482–92.
[15] Roberts MC. Tetracycline therapy: update. Clin Infect Dis 2003;36:462–7.
[16] Stein GE, Craig WA. Tigecycline: a critical review. Clin Infect Dis 2006;43:518–24.
[17] Guh AY, Limbago BM, Kallen AJ. Epidemiology and prevention of carbapenem-resistant Enterobacteriaceae in the United States. Expert Rev Anti Infect Ther 2014;12:565–80.
[18] Michalopoulos A, Falagas ME. Colistin and polymyxin B in critical care. Crit Care Med 2008;24:377–91.
[19] Ayanniyi AA, Ayanniyi RO. A 37 year old woman presenting with impaired visual function during antituberculosis drug therapy: a case report. J Med Case Rep 2011;15:317.
[20] Peloquin CA. Therapeutic drug monitoring in the treatment of tuberculosis. Drugs 2002;62:2169–83.
[21] Babalik A, Babalik A, Mannix S, Francis D, menzies D. Therapeutic drug monitoring in the treatment of active tuberculosis. Can Respir J 2011;18:225–9.
[22] Samuelson J. Why metronidazole is active against both bacteria and parasites. Antimicrob Agents Chemother 1999;43:1533–41.
[23] Klepser M. The value of amphotericin B in the treatment of invasive fungal infections. J Crit Care 2011;26:225. [e1-10].
[24] Balakrishnan I, Shorten RJ. Therapeutic drug monitoring of antimicrobials. Ann Clin Biochem 2016;53:333–46.
[25] Practical handbook of microbiology. Boca Raton, FL, CRC Press. Taylor & Francis Group; 2009.
[26] Antimicrobial susceptibility testing. Boca Raton, FL, CRC Press. Taylor & Francis Group; 2007.
[27] Clinical and Laboratory Standards Institute (CLSI). Performance standards for antimicrobial susceptibility testing. CLSI supplements M100S, 26th ed, 2016.
[28] De Clercq E. Antiviral drugs in current clinical use. J Clin Virol 2004;30:115–33.
[29] Pau AK, George JM. Antiretroviral therapy: current drugs. Infect Dis Clin North Am 2014;28:371–402.

CHAPTER 8

Viruses Including Human Immunodeficiency Virus

Contents

INTRODUCTION

Viral agents are common and do not require an immunocompromised host to cause infection. Viruses are nonliving entities that replicate only when they are inside host cells, often using the molecular machinery

Microbiology and Molecular Diagnosis in Pathology.
DOI: http://dx.doi.org/10.1016/B978-0-12-805351-5.00008-9

of the host genome to replicate. The viral genome consists of either ribonucleic acid (RNA) or deoxyribonucleic acid (DNA) which can be comprised of a single or double strand of either nucleic acid. Viral capsids are polyhedral structures composed of proteins that enclose the nucleic acid of the virus. Viruses can also have an envelope which is derived from phospholipids and proteins of the host cell membranes. Electron microscopy (EM) was once used as a method of identification and some viruses have a distinct morphology which will be discussed in their sections [1,2]. This chapter discusses the following points which are necessary when diagnosing an infection and providing adequate antibiotic treatment:

- Basic physical, chemical, and structural characteristics.
- Epidemiology of each infection.
- Pathophysiology changes associated with infection, including host-related factors.
- Clinical manifestations of infection.
- The biochemical assays useful in differentiating genera from other closely related.

DIAGNOSTIC APPROACH

Viruses are small entities (20–300 nm) that infect and replicate within host cells, using host cellular machinery to synthesize new infectious particles and create progeny virions (virus particles). On one hand, a virus represents an inert biochemical complex of macromolecules, since it cannot replicate outside of a living cell. However, viruses are known to infect all living organisms. Indeed, a broad array of viruses contribute to manifestations of human disease. The extent of infection and related disease state depends upon the number of virions infecting the host, as well as specific host responses that contribute to associated manifestation of pathology. Diagnosis can be made by serology, growth in cell culture, cytopathic effect (CPE) including presence of intranuclear or intracytoplasmic inclusions and immunohistochemical staining of tissue biopsies or molecular methods.

RIBONUCLEIC ACID VIRUSES

These viruses contain only RNA as the genomic content.

Human Immunodeficiency Virus

Human immunodeficiency virus (HIV) is the cause of acquired immunodeficiency syndrome (AIDS). HIV is a single-stranded RNA virus and requires the enzyme reverse transcriptase for replication. Primary infection is considered an asymptomatic phase during which the virus is replicating, followed by an increased viral load and finally immune dysfunction and AIDS. HIV-1 is the predominant strain in the United State, Europe, and Australia while HIV-2 is predominantly found in West African nations. After infection with HIV-1, specific indicators arise in the blood with HIV-1 RNA first, followed by protein p24 antigen, and then antibodies directed toward HIV-1 epitopes.

Replication of the virus requires attachment of glycoproteins to host receptors. The protein gp120 attaches to the host cellular receptor on CD4+ cells. This binding allows for the interaction with the chemokine co-receptor sites CXCR4 and CCR5. Rapid evolution of the virus can cause viral turnover and a high mutation rate in some patients allowing for immune escape. Transmission can include sexual contact through mucosal membranes. Although rare, accidental transmission through needle sticks in laboratories has been noted. Congenital HIV can occur in neonates born to HIV positive mothers, particularly those with uncontrolled HIV.

Screening tests for HIV measure the presence of specific antibody in the patient's serum, detectable after the host has been exposed to the virus for some time (window period). Enzyme immunoassays (EIAs) have developed over time from first to third generation assays. With each new generation the tests have become more sensitive and were associated with a concomitant decrease in the window period. The fourth generation tests detect HIV antibody and the p24 antigen, and the fifth generation additionally differentiates between antibody and antigen allowing assessing the acuteness of the infection and also differentiates between HIV-1 and HIV-2. Both of these new assays decrease the window period almost to the extent of molecular testing. Either an indirect immunofluorescence assay (IFA) or western blot (WB) can be used to confirm detection of antibodies to HIV-1. The WB requires reactivity to at least two of the three following bands p24, gp41, and gp120/160. The confirmatory test for the fourth and fifth generation assays is the multispot test which also differentiates HIV-1 and HIV-2.

Molecular testing is now more accessible and both RNA and DNA testing can be performed. Real-time reverse transcriptase PCR can determine HIV-1 RNA viral load in plasma which is helpful in staging a patient during disease progression and in following response to treatment. Blood should be collected in a tube containing EDTA. DNA testing can be used to identify disease in newborns, since the presence of DNA is an indicator of replication and incorporation of virus in the host cell genome.

The HIV has a tendency to mutate and become resistant to commonly used antivirals, therefore, genotype or phenotype testing is often needed to guide treatment. Genotypic assays employ molecular methods to identify viruses resistant to protease inhibitors, nucleoside reverse transcriptase inhibitors, nonnucleoside reverse transcriptase inhibitors, and integrase inhibitors by sequencing the patient's HIV virus. Phenotypic assays require the growth of the virus in various concentrations of the antiviral drug to determine susceptibility of the virus and are therefore more time consuming and costly.

There are multiple antivirals that are available which target the viral replication including nucleoside analogs, protease inhibitors, and reverse transcriptase inhibitors. Tenofovir disoproxil/emtricitabine is a new antiviral combination drug containing a reverse transcriptase inhibitor and a nucleoside reverse transcriptase inhibitor. This antiviral can treat HIV infection and reduce the risk of HIV transmission (see Chapter 7: Antibiotics, Antimicrobial Resistance, Antibiotic Susceptibility Testing, and Therapeutic Drug Monitoring for Selected Drugs).

Influenza Virus

Orthomyxoviruses are enveloped, single-stranded RNA viruses which have a lipid bilayer containing hemagglutinin (HA) and neuraminidase (NA) glycoproteins. There are currently 16 HA subtypes and 9 NA subtypes. The influenza virus is divided into types A, B, and C, of which in types A and B the HA and NA antigens undergo genetic variation; the C type is generally antigenically stable.

Influenza initially causes upper respiratory tract infections which can spread to the lower respiratory tract and cause severe desquamation of bronchial or alveolar epithelium. Complications of influenza can result in pneumonia or with bacterial pneumonia and possible neurologic syndromes. Transmission occurs through airborne droplets or direct contact with infected patients. The natural hosts of the virus are avian and mammalian species.

Novel strains of influenza occur when HA and NA antigens change to the various subtypes which permits evasion of host immune responses and allows for reinfection. Cell cultures were previously useful for isolation and diagnosis of the influenza virus. Various cells were used to identify CPEs of the influenza virus including primary monkey kidney cells, human diploid lung fibroblasts, 549 human carcinoma cells, and Madin–Darby canine kidney cells. Unfortunately, culture requires several weeks. Presumptive identification of the presence of influenza virus can be made by hemagglutination using chicken red blood cells, although other viruses such as parainfluenza, mumps, and measles produce HAs.

Antigen testing is commonly used, particularly in the out-patient setting or emergency rooms, for the rapid detection of influenza, however, sensitivity is low. Molecular testing is more sensitive and is available in formats that require testing in a molecular lab as well those that can be performed as a point of care test (see Chapter 12: Overview of Molecular Diagnostics Principles). Both antigen tests and molecular methods can differentiate Influenza A from B; however, molecular tests can also identify the most common subtypes by their HA and NA, which is useful for epidemiologic purposes (H1N1 or H3N2).

Treatment for influenzae is oseltamivir which should be administered within 48 hours of illness onset. An inactivated and recombinant influenza vaccine is available as well as a nasal spray which is a live attenuated influenza vaccine. Most hospitals require personnel to be vaccinated to prevent spread to other caregivers and patients.

Respiratory Syncytial Virus and Metapneumovirus

These viruses belong to the *Paramyxoviridae* family and are enveloped, single-stranded RNA viruses. Respiratory syncytial virus (RSV) is the cause of severe acute lower respiratory tract illnesses in children including bronchiolitis, tracheobronchitis, and pneumonia. In adults, patients who are at-risk are those who live in nursing homes, and those with underlying heart and lung conditions. Metapneumovirus is the cause of the common cold and causes lower and upper respiratory tract illnesses. Transmission occurs through airborne droplets or direct contact with infected patients.

Cytopathological effects of RSV include formation of multinucleated giant cells, or syncytia, in cell culture. Cells available for tissue culture include human heteroploid cells such as HEp-2, HeLa, and A549 cells. Rapid antigen tests are available as IFAs, EIA, and chromatographic immunoassay. Molecular methods can determine the presence of the virus and is often included in respiratory multiplex assays.

Parainfluenza and Mumps

The human parainfluenza (HPIV) and mumps virus also belong to the *Paramyxoviridae* family of enveloped viruses. HPIV 1–4 are associated with upper respiratory tract infection in pediatric and adult patients. Disease is self-limiting and reinfection is possible due to the lack of long-term immunity. HPIV is the most common cause of croup in infants to children 6 years of age. Mumps can be asymptomatic in up to 30% of cases. Nonspecific or respiratory symptoms are common and slightly elevated temperature and enlargement of parotid glands are a common characteristic of infection. HPIV and mumps are transmitted via airborne droplets or contact with contaminated fomites.

A live virus vaccine for measles, mumps, and rubella (MMR) is available. Molecular methods are more sensitive and specific for diagnosis of HPIV and can differentiate the types 1–4. Treatment includes ribavirin.

Measles and Rubella

Measles virus is part of the family *Paramyxoviridae* and is highly contagious, transmitted via aerosols. Measles, also known as rubeola, occurs as acute respiratory illness with a prodrome of fever that can reach 105°F with malaise, cough, coryza, and conjunctivitis. Pathognomonic enanthema or Koplik spots are followed by a maculopapular rash. Exanthema will occur on the face and upper body followed by the trunk and upper, then lower extremities. Lesions persist for about 10 days and will finally resolve to fine desquamation. Complications that can occur include otitis media, bronchopneumonia, laryngotracheobronchitis, diarrhea, and acute disseminated encephalomyelitis and subacute sclerosing panencephalitis. Serology testing for measles-specific IgM can be used for diagnosis as well as molecular methods.

Rubella belongs to the *Togaviridae* family and is transmitted through direct contact of nasal and throat secretions and inhalation of droplets. Rubella, also known as German measles, occurs as a mild, maculopapular rash with lymphadenopathy and fever. The rash will start on the face and last about 3 days. Complications that can occur include thrombocytopenic purpura and encephalitis. Infection during pregnancy can lead to congenital defects in infants with shedding of the virus for a long period of time. Prenatal screening for immunity by serum IgG tests is recommended for all pregnant women at the initial prenatal visit. Viral detection can be made from throat, nasal, or urine sample by molecular methods and serology which tests for presence of rubella-specific IgM. A live virus vaccine for MMR is available.

Flaviviridae

The flavivirus family includes West Nile virus, Dengue virus, Yellow fever, St. Louis encephalitis, Chikungunya, and Zika virus. These are considered arboviruses which originated from ARthropod-BOrne virus, and are transmitted via mosquitoes. Flaviviruses are enveloped single-stranded RNA viruses that cause mostly asymptomatic infection. While many infections are asymptomatic, infection may begin as a nonspecific, febrile illness and develop as a severe and life-threatening disease. Since these viruses are not endemic in the community, diagnosis is made by antibody testing (IgG) in the serum. Testing should be limited to the warm season when mosquitoes are viable. Characteristics of various flavivirus are summarized in Table 8.1.

Enterovirus and Parechovirus

These viruses belong to the *Picornaviridae* family and are small, nonenveloped viruses that are comprised of a single-stranded RNA genome. Viruses included in this family are enterovirus, echovirus, parechovirus, poliovirus, and coxsackie virus. Many infections are asymptomatic and disease can be localized or systemic. Neurologic syndromes include meningitis, encephalitis, and acute flaccid paralysis. Respiratory syndromes can include the common cold, hand-foot-and-mouth disease, pharyngitis,

Table 8.1 Characteristics of various flavivirus

Flavivirus	Vector	Location
Dengue virus	*Aedes aegypti*, *Aedes albopictus*	Central America, South America, Caribbean, Southeast Asia, US states Florida, Hawaii, Texas
Yellow fever virus	*Aedes aegypti*	Sub-Saharan Africa, South and Central America, Caribbean islands
Zika virus	*Aedes* spp., *Culex* spp., *Anopheles* spp.	South America, Central America, Mexico, US states Texas and Florida
West Nile virus	*Aedes* spp., *Culex* spp., *Anopheles* spp.	North America, Africa, Europe, Middle East, West Asia
Chikungunya virus	*Aedes* spp., *Culex* spp., *Anopheles* spp.	Africa, South America, Central America, North America, Asia, Europe, Pacific Islands
St. Louis encephalitis virus	*Culex*	United States, Canada, Mexico

Table 8.2 Characteristics of enteroviruses

Enterovirus	Species	Serotypes infecting humans
EV A	Coxsackie viruses Enteroviruses	20
EV B	Coxsackie viruses Echoviruses Enteroviruses	59
EV C	Coxsackie viruses Polioviruses Enteroviruses	23
EV D	Enteroviruses	4, including EV-D68
Rhinovirus	A and B	107
Human parechovirus	HPeV	19

bronchiolitis, rhinitis, and pneumonia as well as cardiovascular such as myocarditis. Poliovirus causes paralytic poliomyelitis, aseptic meningitis, and febrile illness. Echovirus causes aseptic meningitis, rash, febrile illness, conjunctivitis, and severe generalized neonatal disease.

Transmission of the enterovirus is largely by the fecal-oral route, except for the respiratory enterovirus strain, EV-D68, transmitted through airborne droplets. Enteroviruses are more prevalent in the summer and fall months. Rhinovirus can occur year round. Poliovirus is spread through fecal-oral route and is more common in areas with poor sanitation. Eradication efforts have decreased polio cases 106 total cases in 2015 according to the Polio Global Eradication Initiative [3].

Primary monkey kidney cells can be used to grow most enteroviruses in cell culture from cerebrospinal fluid (CSF), blood, respiratory, throat swabs, and stool specimens. Antibody testing is recommended for diagnosis of coxsackie viral myocarditis since not all strains will grow in cell culture. Molecular methods are the preferred method for diagnosis of meningitis and disseminated disease, however, not all commercially available enterovirus PCR assays include parechovirus, mostly seen in neonates. Although supportive treatment is usually used ribavirin can also be used. There is a vaccine against poliovirus, but not the other enteroviruses. Characteristics of enteroviruses are listed in Table 8.2.

Coronavirus

Human corona viruses are endemic in the human population. The causative agent of severe acute respiratory syndrome (SARS) is SARS virus which emerged in the Guangdong province of China. Middle East Respiratory Syndrome (MERS) was initially isolated in Saudi Arabia. Birds and wild mammals are natural reservoirs and can be transmitted as a zoonotic virus via airborne transmission.

Endemic human corona viruses cause upper respiratory tract infections in patients. Symptoms include fever, cough, sore throat, rhinorrhea, and croup in children. Infections occur more often in the winter months. SARS virus was transmitted via airborne transmission and caused the SARS pandemic which spread to 26 countries with about 700 deaths according to the The Center for Disease Control and Prevention (CDC). Symptoms include fever, malaise, nonproductive cough followed by rapid respiratory deterioration. MERS virus also causes illness with high mortality rates especially in patients with comorbidities. Symptoms include fever, cough, pneumonia, and respiratory deterioration. Molecular methods such as RT-PCR are available for detection.

Hepatitis A and Hepatitis E

Hepatitis A (HepA) and Hepatitis E (HepE) viruses are both nonenveloped viruses belonging to different families. HepA is part of the *Picornaviridae* family, while HepE is part of the *Hepeviridae* family. HepA and HepE are commonly seen in waterborne outbreaks associated with contaminated food and transmitted through the fecal-oral route. HepA virus can be excreted in high titers in feces allowing for person to person spread. HepE can be ingested in meat from infected animals, mainly in Western Europe and Japan. Zoonotic infections from pigs and deer can also occur.

HepA and HepE viruses cause viral hepatitis with initial presentation of fever, headache, anorexia, nausea, and abdominal discomfort with more severe disease in older patients. HepA will infect the cells lining the alimentary tract and will replicate in the liver. Serology is available for testing. A vaccine is available for prevention of HepA virus.

Hepatitis C

Hepatitis C (HepC) virus is part of the *Flaviviridae* family and is its own genus, *Hepadnavirus*. HepC is the most common blood borne infections in the United States. Infection can range from asymptomatic carriers, to clearance of virus, and chronic infection. Chronic infection will cause liver damage and leads to cirrhosis and is the leading cause of liver transplantations. HepC is transmitted by intravenous drug use, multiple sexual partners, or contact with an infected partner.

Diagnosis is made by an initial screen using serology and if positive molecular testing for confirmation. Both qualitative and quantitative testing can be done to determine presence of virus and viral load respectively. Treatment can include direct acting antiviral drugs including second generation drugs such as sofosbovir, a viral protein inhibitor, as well as ribavirin. HepC virus genotyping may be important to determine the severity and aggressiveness of infection, as well as for understanding response to antiviral therapy. HepC virus genotypes are separated into six distinct genotypes, 1–6, where genotype 1 is the most common in the United States, but can be the most difficult to treat, while 2 and 3 respond well to ribavirin therapy [4].

Filoviruses

Ebola virus and Marburg viruses belong to the *Filoviridae* family. Symptoms are similar but vary greatly between individual patients. The incubation period can be between 4 and 16 days and resembles a flu-like illness followed by nausea, vomiting, sore throat, and diarrhea. A maculopapular rash will follow with mucosal membrane hemorrhage and gastrointestinal bleeding. Death will occur after shock between 6 and 16 days after onset of illness. Virus detection requires molecular methods in specimen collected during the acute stage of illness. During the most current outbreak of 2014–16, the CDC recommended that patients with suspected ebola virus be tested only in special Laboratory Response Network laboratories located in specific cities [5]. In addition, hospitals were asked to care for these patients in areas in which the staff engaged in testing was limited, equipment used was segregated, and performed in dedicated places for these individuals.

DEOXYRIBONUCLEIC ACID VIRUSES

DNA viruses contain only DNA as part of their genetic makeup. Various DNA viruses are discussed in this section.

Herpes Simplex Virus

Human simplex virus (HSV)-1 and HSV-2 belong to the family *Herpesviridae*. Primary infection for HSV-1 can be acquired in childhood without symptoms and can cause fever or submandibular lymphadenopathy in older children. Primary infection with HSV-2 occurs as herpes genitalis, fever, and lymphadenopathy with healing lesions that typically last approximately 3 weeks.

After primary infection, the virus will become latent in the dorsal root ganglia, recurring periodically. HSV-1 is typically associated with orolabial recurrence while genital recurrence is usually caused by HSV-2. However, genital infections caused by HSV-1 are becoming more common. Neonatal herpes can occur during vaginal delivery. Risk of infection is higher with mothers who have a primary infection. Disease can range from localized infection in the skin, eyes, and mucosa to manifestations that can at times occur as CNS or disseminated disease. HSV can cause central nervous system (CNS) disease in an immunocompromised host. HSV encephalitis can be fatal in about 70% of cases without therapy. With therapy this is reduced to about 20%. HSV meningitis is usually self-limiting and may be seen as sequelae to genital herpes. HSV meningitis may recur in up to 25% of cases after an asymptomatic period of months to years.

Transmission occurs through direct contact with secretions from an infected person. The incidence of asymptomatic carriers is high. Although HSV multiplies rapidly compared to other viruses and can grow in Rabbit Kidney cells in as fast as 24 hours, molecular methods are the most appropriate method for diagnosis of encephalitis or meningitis. EM images are reminiscent of a round wheel but are indistinguishable from the other herpes viruses. Tzanck smears (demonstrating multinucleated giant cells) are still used for diagnosis of fresh, untreated, un-roofed skin lesions, or superficial surfaces in newborns. Therapy should be started on clinical grounds if early PCR is negative. Treatment available includes acyclovir and valacyclovir.

Varicella Zoster Virus

Varicella Zoster virus (VZV) is part of the family of *Herpesviridae*. Varicella, or chicken pox, is the initial disease of VZV. Primary infection occurs in childhood. Incubation time is 10–21 days during which skin vesicles produced are highly infectious. VZV will become latent and remain in the ganglia. Encephalitis can occur during chicken pox and occurs with acute cerebellar ataxia. Reactivation will occur in the ganglion and tracks down the sensory nerve to the area of the skin innervated by the nerve forming a rash. VZV is acquired by airborne transmission. Latent VZV infection establishes in the dorsal root ganglia. Although clinically different from HSV due to the presence of a dermatomal rash, the Tzanck smear described for HSV, can also be used for VZV, however, differentiation of the two agents would not be possible using this technique. Treatment includes acyclovir.

Cytomegalovirus

Cytomegalovirus (CMV) is in the family *Herpesviridae*, and was formally called HHV-5. CMV can be asymptomatic but is the most common viral cause of congenital defects. Congenital defects include microcephaly, intracerebral calcification, hepatosplenomegaly, and cytomegalic inclusion disease. The risk for severe birth defects is highest when mother has primary infection during pregnancy. CMV is an opportunistic pathogen and will cause pneumonitis, pneumonia, and mortality if not treated. CNS or GI disease as well as disseminated disease can occur in immunocompromised patients.

Transmission is through sexual contact, or via any body fluids that allow transfer by direct contact with the infected person. The fetus can be infected via mother's viremia or through viral ascension from the cervix.

Cell culture can be done in human fibroblast cells. The cytopathogenic effect of CMV results in intranuclear inclusions usually identified by antibody staining. In tissue biopsy these inclusions are called the "owl's-eye." However, due to the slow growth of the virus in tissue culture this method of diagnosis has fallen out of favor. Serology can be useful for testing the immune status of pregnant women and pretransplant patients, however, molecular methods can detect virus in body fluids including CSF and quantitation in plasma is useful in following treatment. Treatment can include ganciclovir, valganciclovir, foscarnet, and cidofovir. Genotypic resistance testing can identify the presence of nucleic acid mutations in the genes *UL97* and *UL54*. Mutations in *UL97* encode resistance to ganciclovir, while mutations in *UL54* encode resistance to ganciclovir, cidofovir, and foscarnet.

Human Herpes Virus 6–8

Human herpes virus (HHV)-6 and HHV-7 belong to the *Roseolovirus* genus. HHV-8 is part of the *Rhadinovirus* genus. HHV-6 is associated with the acute phase exanthema subitum, known as roseola. Symptoms are more severe in immunocompromised patients. Symptoms of primary infection include fever, rhinorrhea, diarrhea, and roseola. In immunocompromised patients, HHV-6 can reactivate and cause encephalitis, pneumonitis, graft-versus-host disease, and bone marrow suppression. HHV-7 can cause roseola, febrile illness, and seizures in children. Reactivation is possible in immunocompromised patients. HHV-8 is associated with all forms of Kaposi's sarcoma, an angioproliferative disease which occurs as brown lesions on the skin, internal organs, and extremities.

HHV-6 and HHV-7 are both found in the saliva and therefore airborne transmission is a likely route of infection. Reactivation of both agents is possible in immunocompromised patients. Greater than 90% of the population is seropositive for HHV-6 and the virus can be transmitted to the fetus during pregnancy. HHV-7 can also be found in the saliva, but congenital disease has not been recorded. HHV-8 can be found in the saliva, breast milk, and semen. Transmission can occur through injection drug use and blood transfusion.

Molecular methods are the appropriate tests for these viruses. CPEs in cell culture occurs as multinucleated giant cells. Treatment includes antivirals targeting HHV-6 and HHV-7 which include foscarnet, cidofovir, and ganciclovir.

Hepatitis B and Hepatitis D

Hepatitis B (HepB) is a part of the *Hepadnaviridae* family. HepB infects hepatocytes with disease ranging from asymptomatic to acute infection that resolves to a chronic infection. There are three phases of chronic Hepatitis B Virus (HBV) disease: the immune tolerant phase, the immune clearance phase, and the inactive carrier phase. Hepatitis D requires HepB to replicate and is only transmitted to patients already infected with HepB. The most common modes of transmission are from mother to child, early childhood infections from close contact, sexual contact, and transfer in shared needles during intravenous drug use. Assays are available to test for HepB proteins and specific antibodies in serum. Quantitative molecular methods are commonly used once diagnosis is made.

Various proteins can be detected during the course of infection. HepB surface antigen (HBsAg) and the HepB envelope antigen (HBeAg) are present in the serum during primary infection. The presence of HBsAg is useful in diagnosis. The presence of HBeAg can indicates viral replication producing a highly infectious patient. Serology testing for HepB-specific antibodies is used to determine the stages of diseases and establish the immunity after vaccination. The first IgM antibodies to appear are anti-Hepatitis B core (anti-HBc) which can persist for weeks after initial infection. The presence of anti-HBs indicates immunity from infection or vaccination. As acute infection resolves, anti-HBe is detected and HBsAg disappears. Persistence of HBsAg indicates a carrier state or, if symptoms are present, a chronically infected patient. Nearly 80% of primary liver cancer (Hepatocellular carcinoma) is associated with chronic

HBV infection. Treatment includes lamivudine, adefovir, tenofovir, and telbivudine. In addition a subunit vaccine derived from HBsAg for HeB is available.

Adenovirus

Adenoviruses belong to the *Adenoviridae* family. Infection can be asymptomatic and cause respiratory infections and diarrheal illness. Symptoms include fever, pharyngitis, and cervical adenopathy. Other infections include involvement of the eye, gastrointestinal tract, urinary tract, CNS, liver, and genital tract. Gastroenteritis is normally caused by the adenovirus serotype 40/41. Transmission includes airborne transmission and fecal-oral route.

Adenovirus can grow in tissue culture in human heteroploid cells such as Hep-2. CPEs are composed of aggregates of refractile cells. Molecular methods are available for testing and can be done on serum, CSF, respiratory, and stool samples. Treatment can include ganciclovir, ribavirin, and cidofovir.

Papovavirus

Double-stranded DNA viruses also include the papillomaviruses and the polyomaviruses.

Human Papillomaviruses

Multiple papillomaviruses have an oncogenic potential, especially the high risk serotypes HPV-16 and HPV-18. Hepatitis B Virus (HPV) infects and replicates in the squamous epithelium and mucosal membranes which causes benign warts and papillomas. HPV can cause oncogenic cervical cancer. Transmission occurs through direct contact of infected epithelia especially through sexual contact.

Cytology of the exfoliated cervical epithelial cells shows characteristic features using Papanicolaou stain. CPEs are characterized by high nucleus/cytoplasm ratios. Molecular testing can be used for diagnosis and to specifically identify the high risk serotypes. A vaccine is available for HPV Gardasil (Merck, West Point, PA) which targets HPV types 6, 11, 16, and 18.

Polyomavirus

Polyomaviruses are part of the *Polyomaviridae* family. BK virus establishes latent infection in the kidney and is shed in the urine. Disseminated disease usually occurs in immunocompromised patients, BK virus causes renal disease in kidney transplant patients and JC virus causes neurologic symptoms and progressive multifocal leukoencephalopathy in HIV infected patients, both with poor prognosis.

Infectious spread likely occurs via airborne transmission through direct contact of bodily fluids; although spread can also occur through organ transplantation or through occurrence as congenital infections. Urine cytology can identify large cells in the urine (decoy cells) however diagnosis is usually made by real-time PCR of urine and/or blood. JC virus can be detected in the CSF by molecular methods.

Rabies Virus

Rabies belongs to the genus *Lyssavirus* and is a zoonotic disease which causes acute, fatal viral encephalomyelitis. Transmission occurs from small rodents that vary based on geography. In the United States, rabies can be transferred from skunks in California and North and South Central states, gray foxes in Texas and Arizona, raccoons on the East coast, Arctic and red foxes in Alaska, and mongooses in Puerto Rico.

The rabies virus infects the CNS and occurs with fever, headache, and weakness. Progression of the disease results in insomnia, anxiety, confusion, paralysis, hallucinations, agitation, hypersalivation, and hydrophobia. Death typically occurs within days of onset of these symptoms. Diagnosis requires saliva, serum, CSF and a skin biopsy of hair follicles at the nape of the neck. The serum and saliva are tested for antibodies to the rabies virus, while the skin biopsy is examined for rabies antigen. Brain biopsy of the infected animal may demonstrate Rabies inclusions or Negri bodies. EM images will show morphology reminiscent of a bullet. Characteristics of DNA viruses and RNA viruses are summarized in Table 8.3.

KEY POINTS

- Viruses are nonliving entities that replicate inside host cells, using the molecular machinery of the host to replicate single or double-stranded RNA or DNA. Glycoproteins present on the viral capsid surface interact with specific host membrane molecules, allowing adhesion and subsequent infection.
- RNA viruses, such as the HIV that causes AIDS, may be found both as free virus particles and as virions within infected host cells.
- The orthomyxoviruses are enveloped, single-stranded RNA viruses which have a lipid bilayer containing HA and NA glycoproteins. The influenza virus is a prime example, with surface antigens that change under pressure of immune recognition.

Table 8.3 Characteristics of viruses

Family	Genome	Envelope	Cell line	Cytopathic effect
Deoxyribonucleic acid viruses				
Herpesviridae HSV,[a] VZV, CMV, EBV,[b] HHV-6–8[b]	dsDNA	+	HDF	Nuclear inclusions (Cowdry)
Adenoviridae	dsDNA	–	Hep-2	Grape-like clusters
Papillomaviridae	dsDNA	–	–	–
Poxviridae Small pox, Molluscum	dsDNA	–	–	–
Polyomaviridae BK, JC	dsDNA	–	–	Decoy cells
Ribonucleic acid viruses				
Orthomyxoviridae Influenza	ssRNA	+	PMK	Hemadsorption
Paramyxoviridae Measles,[c] Mumps, Parainfluenza	ssRNA	–	PMK	Cytoplasmic inclusions, syncytia
Picornaviridae Entero, Coxsackie,[d] Rhino	ssRNA	–	PMK	Tear-shaped cells
Rhabdoviridae Rabies	ssRNA	+		Negri bodies, bullet shaped
Togaviridae Arbo, Rubella	ssRNA	+	–	
Retroviridae HIV	ssRNA	+	–	
Reoviridae Rotavirus	dsRNA	–	–	

[a]Rapid rounding of all cells.
[b]No growth cell culture.
[c]Measles nuclear and cytoplasmic inclusions, Warthin–Finkeldy cells.
[d]Focal swollen cells; *HDF*, human diploid fibroblast; *PMK*, primary monkey kidney cells; *Hep-2*, human epithelial type 2.

- The arbovirus flavivirus family includes West Nile virus, Dengue virus, Yellow Fever St. Louis encephalitis, and Zika virus. These are transmitted via arthropod vectors.
- The *Paramyxoviridae* family of enveloped viruses includes the parainfluenza, mumps, measles, rubella, and RSV. Transmission typically occurs through airborne droplets or direct contact with infected patients.
- The *Picornaviridae* family is small, nonenveloped single-stranded RNA viruses which include enterovirus, echovirus, parechovirus, poliovirus, and coxsackie virus. Clinical symptoms postinfection range from respiratory disorder to rashes.
- Corona viruses, represented by the agent causing SARS, are zoonotic viruses that can be spread via airborne transmission.
- The Hepatitis viruses represent a broad group of agents that span multiple families. All replicate in the liver, with resultant typical chronic infection.

- DNA viruses contain only DNA. A common member of this class is the *Herpesviridae* family, which includes the HSV agents.
- Other DNA viruses include those of the *Hepadnaviridae* family (Hepatitis) and the *Herpesviridae* family (Varicella Zoster and CMV). Transmission is acquired through contact with body fluids, typically via sexual or other means.
- The Roseolovirus HHVs typically cause clinical exanthems, seen due to CPEs postinfection of host cells.
- Other DNA viruses of clinical note include [1] adenoviruses (*Adenoviridae* family) that lead to respiratory infections and can cause diarrheal illness; [2] human papillomaviruses with oncogenic capabilities; and [3] polyomavirus (*Polyomaviridae*) which cause renal and neurological disorders.
- Molecular diagnostics allows rapid detection by matching of nucleic acid sequences, by detection of viral antigens released during the infection process, or by detection of increased antibody presence with specificity for viral proteins.
- Common antiviral therapeutics function via direct inhibition of reverse transcriptase. Alternative antiviral agents utilize nucleoside analogs to disrupt DNA polymerase activity to limit virion replication.

REFERENCES

[1] Richman DD, Whitley RJ, Hayden FG. Clinical virology. 3rd ed. Washington, DC: American Society for Microbiology; 2009.
[2] Jorgensen JH, Pfaller MA. Manual of clinical microbiology. 11th ed. vol. 1. 2015. ASM Press, Washington, D.C.
[3] WHO. Poleomyeltis fact sheet. Retrieved from http://www.who.int/immunization/diseases/poliomyelitis/en/; April 2016.
[4] Zein NN. Clinical significance of hepatitis C virus genotypes. Clin Microbiol Rev 2000;13:223–35.
[5] Karwowski MP, et al. Clinical inquiries regarding ebola virus disease received by CDC—United States, July 9–November 15, 2014. MMWR Morb Mortal Wkly Rep 2014;63(Early Release):1–5.

CHAPTER 9

Overview of Fungal Infections

Contents

INTRODUCTION

Systemic fungal infections have become increasingly common worldwide [1]. This is likely due to the extended lifespan of critically ill patients even in the face of discovery of newer antifungal agents [2]. Immunocompromised individuals and those on immunosuppressive compounds are considered at-risk populations for fungal infections. Immunocompetent individuals with traumatic implantation of organic material in environmental settings have also been seen to acquire fungal infections [3]. Fungi can be seen directly in tissue with various stains and presumptively categorized based on the presence and appearance of their hyphae including presence of septations, angle of branching, and their pigmentation. However, specific identification can only be made by growth of the fungus in culture. While biochemical identification can be used for the identification of yeast, macroscopic appearance of the colony along with microscopic examination using lactophenol cotton blue staining

Microbiology and Molecular Diagnosis in Pathology.
DOI: http://dx.doi.org/10.1016/B978-0-12-805351-5.00009-0

of a tease or tape prep is the primary mode of identification for filamentous fungi based on the type of sporulation that is evident. Recently, molecular methods, especially 18S ribosomal RNA, are being used for fungal identification. This chapter discusses the following points which are necessary when diagnosing a fungal infection and providing adequate antifungal therapy.

- Basic physical, chemical, and structural characteristics.
- Epidemiology of each infection.
- Pathophysiological changes associated with infection, including host-related factors.
- Clinical manifestations of infection.
- The biochemical assays useful in differentiating genera from other closely related.
- Most commonly used antifungal treatments.

DIAGNOSTIC APPROACH

Fungi are filamentous, spore producing eukaryotic organisms that do not contain chlorophyll. The mycotic species include molds, yeasts, and higher fungi. Many species grow as saprophytes (requiring organic material for energy) with beneficial utility, however, multiple species are pathogenic. Clinically, the majority of mycotic species can lead to severe infection within the immunocompromised host. Many of the filamentous fungi also elicit toxic effects (mycotoxicosis) upon ingestion, and are known to cause allergic reactions. These agents are classified according to types of tissue colonized, and according to development of clinical disease. Specifically, superficial mycoses are limited to the skin and hair, cutaneous mycoses (including dermatophytes) infect keratinized tissue, subcutaneous mycoses are invasive and infect subcutaneous fascia, muscle, and deeper epidermal layers, and systemic mycoses result from inhalation of spores [4,5].

Yeast/Yeast-Like

Yeasts or yeast-like organisms are eukaryotic, unicellular organisms that reproduce by budding. The budding cells are known as blastoconidia. Some yeast will elongate from the base, or parent cell, and look like hyphal elements which are called pseudohyphae. Chlamydospores are the resting spore form of *Candida*, which survive in unfavorable conditions. The most commonly isolated yeasts are common colonizers of humans and correlation with clinical findings is critical in determining the need for treatment. Yeast/yeast-like characteristics are summarized in Table 9.1.

Table 9.1 Yeast/Yeast-Like Characteristics

Organism	Germ Tube	Blastoconidia	Pseudohyphae	Capsule	Chlamydospore	Antifungal Susceptibility
Candida albicans	+	3–4 µm	+	–	+ (single)	>95% S to fluconazole
Candida glabrata	–	3–4 µm	–	–	–	Most R to fluconazole
Candida lusitaniae	–	3–4 µm	+, curved	–	–	possible R to amphotericin
Candida dubliniensis	–	3–4 µm	+	–	+ (abundant cluster)	Some R to fluconazole
Candida krusei	–	–	+, elongated	–	–	Inherent R to fluconazole
Candida tropicalis	–	3–5 µm	+	–	+	Most S to fluconazole
Cryptococcus spp.	–	3.5–8 µm	Rare	+	–	Echinocandins are ineffective

– = absent, + = present, *S* = susceptible, *R* = resistant.

Candida

The most common *Candida* causing human disease is *Candida albicans*. Other medically important *Candida* species include *Candida parapsilosis* complex, *Candida glabrata* complex, and *Candida dubliniensis*. The yeast are single cells that reproduce by budding cells called blastoconidia, measuring between 3 and 4 µm. The elongated blastoconidia can form chains called pseudohyphae. Colonies on Sabouraud dextrose agar (SDA) are smooth, white to cream. The *Candida* spp. are ubiquitously found in humans as normal flora and the environment. Candidemia is a common cause of hospital acquired bloodstream infection [6], while superficial mucosal disease like vaginal and oral candidiasis and disseminated disease is seen more commonly in immunocompromised patients. Common mode of transmission is direct person to person contact particularly in the hospital setting.

Although *Candida* will stain with a Gram stain, they can be rapidly identified using a wet mount which can show budding yeast. *Candida* can be presumptively identified using a rapid germ tube test. The germ tube test is a diagnostic test in which a sample of yeast cells is suspended in bovine serum and incubated at 37°C for 3 hours. The germ tube is the outgrowth from a conidia or spore. The germ tube test can be performed to differentiate *C. albicans* from the other *Candida* sp. If the yeast is *C. albicans*, the germ tube will be present other species do not produce a germ tube with one exception of *C. dubliniensis*. Colonies of *C. albicans* also appear as a sun with radiations outward or often called feet. *C. albicans* are typically susceptible to azole antifungals including fluconazole as long as the patient has not previously been on treatment. *C. glabrata* is known to have elevated minimum inhibitory concentrations (MICs) to fluconazole and *Candida krusei* are always considered resistant to fluconazole therefore, susceptibility testing of the other *Candida* species is important to determine if azoles can be used. *Candida* species are usually susceptible to the echinocandins, such as micafungin or caspofungin; however, *C. parapsilosis* can have higher MICs than these agents. *Candida* species are also almost always susceptible to amphotericin. The only species that can rarely be resistant to amphotericin is *Candida lusitaniae*.

Cryptococcus spp.

Cryptococcus neoformans subspecies *neoformans* and *Cryptococcus gattii* subspecies *gattii* are the human pathogens of this genus. *Cryptococcus* can cause meningoencephalitis, especially in AIDS patients. Transmission can occur through inhalation of the organisms in the environment and in areas in which bird droppings have accumulated over time. Inhalation of the yeast can spread from the lungs to the brain and meninges. Yeast can be found in cerebrospinal fluid (CSF) as well as blood cultures. On cornmeal tween-80 agar, *C. neoformans* produces round, budding yeast cells. No true hyphae are visible and pseudohyphae are absent or rudimentary. *C. neoformans* can grow at 37°C, while most nonpathogenic species of *Cryptococcus* do not. India ink is used to examine CSF for *Cryptococcus* species. The capsules appear transparent by bright field microscopy and cause a halo effect. Birdseed agar can be used to identify *Cryptococcus* by its ability to produce phenoloxidase. Birdseed agar contains caffeic acid which will serve as a substrate for phenoloxidase and will produce melanin causing the yeast pigmentation to turn brown. *Cryptococcus* is urea positive and can be differentiated from *Rhodotorula*, a yeast that is also urea positive but produces a red pigmented colonies on SDA.

Geotrichum

Geotrichum spp. are opportunistic pathogens that are normally found in the environment as well as part of the normal microbiota of humans. Infection can cause pulmonary, bronchial, oral and vaginal, GI tract, and cutaneous geotrichosis. Yeast forms produce septate hyphae that separate to form chains of individual arthrospores that are rectangular with flat or rounded ends. These arthroconidia are differentiated from those of *Coccidioides* by the absence of disjuncter cells in between the arthroconidia.

Trichosporon

Trichosporon spp. are opportunistic pathogens that can be found on the skin as normal microbiota. Infection can cause white piedra, which occurs as nodules on the hairs of the scalp. *Trichosporon* yeast forms will produce septate hyphae that separate to form elongated arthroconidia. *Trichosporon* spp. are urease positive. They are the only species of yeast that has rarely been known to cause a false positive cryptococcal antigen test.

Saccharomyces

Saccharomyces cerevisiae used in the beer making process can be colonizer or cause of disseminated infections in humans. Ascospores are the sexually produced fungal spore formed within an ascus, produced after growth on ascospore agar. These can be visualized using a Kinyoun acid fast stain.

Malassezia furfur

Malassezia spp. are opportunistic pathogens that can be found on the skin as normal microbiota. Disease can occur as hypopigmentation or hyperpigmentation on the trunk of humans, in addition to causing dandruff and seborrheic dermatitis. *Malassezia* requires lipids to grow and must be incubated with olive oil in culture to grow. Disseminated disease can occur in neonates on total parenteral nutrition due to the presence of lipids. Yeast forms are round to ellipse shaped with one round end and the other a blunt end. The blunt end serves as the location of budding from the phialides. The combination of conidia and small hyphae has an appearance that resembles "spaghetti and meatballs."

THERMALLY DIMORPHIC FUNGI

The thermally dimorphic fungi grow in a yeast-like manner in human tissue or when cultured on enriched medium at 37°C, and as a mold when grown in ambient temperature 25–30°C. The mold form or conidia which are found in the soil are the infective form and convert to the yeast form in the body which is not infectious and is not transmitted person to person. These fungi will grow as a white fuzzy mold at room temperature and can be distinguished by their specific microscopic characteristics. Diagnosis can be made by visualization of the yeast form directly in the specimen or in the tissue or by microscopic appearance of the mold form after growth in the laboratory. DNA probes that can be used on the mold colony are also available for confirmation of identification. Antigen detection in the urine has increased sensitivity for disseminated disease, although detection of antibodies with consistent symptoms can also provide diagnostic value. Antibody testing can be nonspecific and cross reaction with other fungal infections as well as other granulomatous diseases can occur.

Blastomyces dermatitidis

The majority of cases of blastomycosis is seen on the east coast of North America, especially surrounding the Mississippi and Ohio River valley, and Canada, but has been found in Central and South America and Africa. Blastomycosis can range from a pulmonary infection, as transmission occurs through inhalation of conidia to disseminated disease. Extrapulmonary infections include skin lesions and osteomyelitis with skin as the most prominent infection. Yeast can be identified in sputum, tissue, and exudates directly by fungal stain such as Calcofluor white or in tissue by histopathological examination and staining with periodic acid Sciff (PAS) or Grocott-methenamine-silver (GMS), where they appear as broad-based budding cells. *Blastomyces dermatitidis* grows as a white mold at room temperature and has oval microconidia on short stalks that resemble lollipops. Treatment can include amphotericin B, fluconazole, itraconazole, posaconazole, and voriconazole.

Coccidioides immitis/posadasii

Coccidioides immitis is found in northern Mexico and the San Joaquin Valley in California and *Coccidioides posadasii* can be found in the Southwestern states of the United States, including Arizona and Texas. Primary infection has been referred to as valley fever and includes cough, fever, and chills that can last up to a few months. Patients may develop mild, diffuse erythematous rash on the truck and limbs. Dissemination can occur in immunocompromised patients particularly to the central nervous system (CNS). Transmission occurs through inhalation of the arthroconidia from the environment typically following earthquakes or other soil-disturbing events. *Coccidioides* grow as spherules at body temperature which can be seen in tissue by histopathological examination with PAS, GMS, or H&E. The spherules are spherical structures which vary in size and can be as large as 30–60 μm and may contain hundreds of endospores. The spherules can also be broken with release of the endospores which can be seen free in the tissue. *Coccidioides* produce chains of alternating arthroconidia with smaller disjuncter cells in between when grown at room temperature. Unlike the other dimorphic fungi which take several weeks to grow in culture, *Coccidioides* can grow within 2–3 days on routine media and therefore care must be taken in the laboratory not to inhale the arthroconidia, which are very highly contagious. Treatment can include amphotericin B, itraconazole, posaconazole, and voriconazole.

Histoplasma capsulatum

Histoplasmosis is acquired by inhalation of microconidia from the environment particularly areas where there is a disturbance of accumulation of bird or bat guano such as while spelunking in caves. These fungi are usually seen along the Ohio River valley although they can be seen in many areas of North and South America. Histoplasma causes disease in the form of flu-like symptoms. Disseminated histoplasmosis can occur as a chronic illness or fatal disease in patients who are immunocompromised.

The yeast forms can often be seen in a direct smear from the bone marrow or a buffy coat prepared from the peripheral blood in immunocompromised patients with a high load of organisms. The yeast form is small and ranges in size from 1 to 5 μm. They are typically seen within neutrophils or macrophages and appear more uniform in size as compared to Cryptococcus. Serology can be more useful in diagnosis of histoplasmosis as compared to other fungal infections. Antigen testing of the urine is more sensitive in diagnosis of disseminated disease. Complement fixation, which detects both the yeast and mycelial form, is more sensitive but less specific than immunodiffusion. In the immunodiffusion test, two bands can be detected: The M band appears sooner and persists while the H band which is detected mostly during disseminated disease. When grown in culture *Histoplasma capsulatum* produces two types of spores, the tuberculate macroconidia, which range from 8 to 15 μm and microconidia which range from 2 to 5 μm. Treatment includes the antifungals amphotericin B, fluconazole, itraconazole, posaconazole, and voriconazole.

Paracoccidioides brasiliensis

The *Paracoccidioides brasiliensis* fungus is found in areas of Mexico and south to Argentina, with most cases in Brazil, Colombia, and Venezuela. Transmission can occur though inhalation of conidia. Primary infection is in the lungs, but spread to lymph nodes is common. Dormant infection can take years to manifest after primary infection. Chronic disease has substantial lung involvement. Ulcerative mucocutaneous lesions can occur on the face, mouth, and nose. *Paracoccidioides* require a long incubation for growth on rich media. They produce septate hyphae and chlamydospores and produce microconidia. The yeast form has a larger mother cell surrounded by smaller buds, which looks like a ship's pilot wheel. Treatment can include the antifungals amphotericin B, fluconazole, itraconazole, posaconazole, and voriconazole.

Talaromyces marneffei (previously *Penicillium marneffei*)

Talaromyces marneffei is an opportunistic pathogen found in immunocompromised patients, especially those with HIV. Disease can include disseminated infection as well as skin lesions. Transmission occurs after inhalation of spores and is endemic in tropical Asia, including Thailand, Vietnam, and Taiwan. The yeast forms are hyphal elements that produce single celled round conidia that will reproduce by fission with a clear septum and resemble histoplasmosis when seen in a blood smear or bone marrow. The colony will grow as a tan flat colony and will turn a reddish color with a white edge, while the reddish pigment will diffuse into the medium. The hyphal elements appear as conidiophores with terminal metulae, the structure that supports 4–6 phialides, the cell that produces the conidia. This type of microscopic appearance is indistinguishable from other *Penicillium* species which are a common contaminant of bread and human clinical specimens as well.

Sporothrix schenckii

Sporothrix schenckii begins as a skin lesion typically on the extremity with development of nodules that track up the lymphatics and are associated with chronic infection. Exposure is associated with gardeners and those who work with plant material that contain thorns. Transmission is via traumatic implantation. The yeast form is budding cells, sometimes called cigar-shaped bodies. Biopsy of the nodules usually reveals granulomas but no yeast forms. Hyphae are thin with septate and branching conidiophores (hyphae or filaments) that produce small tear-drop shape conidia which form a "rosette" or daisy-like shape. Treatment is with iodine for topical indications or with oral or intravenous itraconazole for disseminated disease.

Emmonsia spp.

Emmonsia spp. are opportunistic pathogens that can cause widespread skin lesions. The yeast forms are thin-walled globose (spherical) to oval with single or multiple polar buds. Hyphae are septate, with numerous smooth-walled conidia on thin pedicles. The terminal conidia establish a floret of 4–8 conidia. Characteristics of various dimorphic fungi are listed in Table 9.2.

THERMALLY NOMOMORPHIC ORGANISMS

Thermally monomorphic organisms are discussed in this section.

Zygomyces

These organisms are environmental filamentous fungi usually found around decaying organic material. Their hyphae are very broad ribbon like and have no or rare hyphal septations, and branch at 90° angles. Sporangiophores contain sporangia which have sporangiospores, the infectious component. These fungi are sometimes found in small outbreaks, including nosocomial infections and in skin lesions after traumatic motor vehicle accidents. They are very fast growing and rapidly destroy tissue, especially in patients with uncontrolled diabetes.

Table 9.2 Thermally dimorphic fungi

Organism	Growth on Sabouraud at 25°C	Morphology at 37°C from tissue or media	Geographical occurrence
Blastomyces dermatitidis	Floccose white mold, oval microconidia, lollipop-like stalks with conidia on ends	Thick-walled yeast with broad-based budding	Eastern North American, N. Ontario, SE Manitoba, Quebec, US Appalachian mountains, W. Lake Michigan, Wisconsin, Mississippi Valley
Coccidioides immitis	Floccose white or gray mold, septate hyphae, barrel-shaped arthroconidia	Spherules with endospores	Western Hemisphere, SW United States
Histoplasma capsulatum	Floccose white or yellow mold, microconidia, and tuberculate macroconidia	Yeast, small budding cells	Ohio River Valley, lower Mississippi River
Paracoccidioides brasiliensis	White and compact mold, microconidia	Yeast, ship pilot wheel	Brazil, Colombia, Venezuela
Talaromyces marneffei	Flat, powdery to velvety, tan becoming reddish yellow with a yellow, white edge/gray-green in the center, broom-like structure as *Penicillium* spp.	Soft, white to tan, dry and yeast-like cells that resemble histoplasmosis	Burma, Cambodia, S China, Indonesia, Laos, Malaysia, Thailand, Vietnam
Sporothrix schenckii	Septate hyphae, turning white to black with age, with delicate conidiophores bearing pyriform (pear-shaped) conidia in rosette clusters (daisy)	Elongated, cigar-shaped yeasts; rarely seen in histological sections of tissue	Ubiquitous
Emmonsia spp.	Septate, hyaline hyphae with numerous smooth-walled conidia on thin pedicles. Terminal conidia establish a floret of 4 or 8 conidia	Yeast thin-walled, globose to oval, with single or multiple polar buds	Ubiquitous

Rhizopus oryzae

Rhizopus can cause localized and disseminated mucormycosis. Sinusitis and pneumonia are the most common types of infection with dissemination prevalent in patients with underlying disease. Inhalation of spores, in addition to traumatic implantation, can cause disease. Growth is apparent on Sabouraud dextrose agar at 25–30°C sometimes after only 24 hours. *Rhizopus* spp. have unbranched sporangiophores with rhizoids, root-like structures, directly under the sporangiophore which can be used to differentiate them from *Mucor* spp. Treatment includes amphotericin B.

Mucor

Mucormycosis can be localized and cause disseminated disease. Sinusitis and pneumonia are the most common with dissemination prevalent in patients with underlying disease. Inhalation of spores will cause infection and have also been seen to invade via traumatic implantation. On Sabouraud dextrose agar at 25–30°C, *Mucor* spp. have branching sporangiophores without rhizoids which can be useful for differentiation from *Rhizopus* spp. Mucorales in the family of Zygomycetes are listed in Table 9.3.

Table 9.3 *Mucorales* in the family of Zygomycetes

Organism	Colony morphology	Sporangiophores	Stolon	Rhizoids
Rhizopus spp.	White, then gray, dense fluffy growth, reverse is gray or brown	Long, rare septa, with dark sporangium at terminal end	Found underneath sporangiophores	Produced where stolons and sporangiophores connect
Mucor spp.	White, then gray, dense fluffy growth, reverse is white	Long, broad nonseptate, branched with round sporangia	None	None
Rhizomucor	Gray, dark brown, reverse is white	Long, nonseptate apical branching, brown	None	Few, short

Rhizomucor

Rhizomucor spp. are opportunistic pathogens. In culture, *Rhizomucor pusillus* is a thermophilic species and can grow in temperatures between 54°C and 58°C. A different species, *Rhizomucor variabilis* has a maximum growth temperature of 38°C. The morphology appears as an intermediate between *Rhizopus* and *Mucor* whereas the sporangiophores are branched and rhizoids can appear in random locations.

HYALINE MOLDS

These molds are opportunistic pathogens of which hyphae appear as colorless, or hyaline, on agar medium. These molds are ubiquitous in the environment and disease varies depending on state of the host. The infectious component is the conidia which can be inhaled.

Aspergillus

Aspergillus spp. can cause pulmonary and respiratory disease as well as cutaneous disease, while invasive disease can occur in severely immunocompromised patients. Diagnosis of disseminated disease can be made using the Aspergillus galactomannan test although it can be positive in other mold infections. Inhalation of spores will allow for transmission of the molds. In tissue specimens the mold will appear as septate hyphae which branch at 45° angles. The mold will grow on Sabouraud dextrose agar and have different colors depending on the media they are grown on and the species. The conidiophores support the phialides which produces the conidia. The phialides can be uniseriate or biseriate meaning that the phialide is directly from the vesicle which is supported by the condiophore or there is a metula supporting the phialade. Treatment can include voriconazole and amphotericin B. Antifungal susceptibility testing is required as resistance has been seen to amphotericin especially in the *Aspergillus terreus* species. Characteristics of *Aspergillus* species are listed in Table 9.4.

Fusarium

Fusarium can cause eye infections, sinusitis, or systemic disease in some patients. Spores can be ingested in contaminated grain or inhaled, however, it is most commonly acquired by traumatic implantation of plant material. *Fusarium* can be recovered from blood, nails, and skin biopsies. Lactophenol blue stain can be used to identify macroconidia and microconidia. Macroconidia are sickle or canoe shaped and contain from several

Table 9.4 *Aspergillus* species characteristics

Organism	Colony morphology	Conidiophore	Phialides
Aspergillus fumigatus	White, then gray or green, reverse is white or tan	Short, smooth	Uniseriate, on upper vesicle
Aspergillus niger	White or yellow, then black, reverse white or yellow	Long, smooth	Biseriate, cover entire vesicle, form "radiate" head
Aspergillus flavus	Yellow or green, reverse red brown	Medium, rough	Uniseriate and biseriate, cover entire vesicle
Aspergillus terreus	Brown, reverse yellow to brown	Short, smooth	Biseriate, columnar

to many compartments. *Fusarium* may have high MICs to amphotericin B and be difficult to treat when disseminated.

Paecilomyces

Paecilomyces can cause cutaneous and subcutaneous infections, sinusitis, and fungemia. Infection can spread through inhalation of conidia or through traumatic implantation. Mold can be recovered from blood and biopsies of the infection site. Lactophenol blue stain can be used to identify the phialides which are longer and tapered than those of *Penicillium*, and establish a single chain on which round conidia are formed. Voriconazole and posaconazole can be used for treatment; however testing is required as various patterns have been recorded.

Dematiaceous

The dematiaceous fungi contain melanin in their cell wall which causes the dark, brown pigmentation of the hyphae. Melanin is resistant to factors the host cell uses to defend itself such as the oxidative burst by phagocytic cells, hydrolytic enzymes, and antifungal [7]. Members of the dematiaceous group are ubiquitous and found in the environment. They can be inhaled, acquired through traumatic implantation, and are occasionally seen as opportunistic pathogens. Phaeohyphomycosis is a broad term that indicates an infection with dematiaceous fungi and is usually a subcutaneous infection, but can disseminate as systemic disease and cause infection in the CNS. Chromoblastomycosis indicates subcutaneous mycoses and disease is characterized by excessive proliferation of skin layers, usually found in tropical and subtropical areas.

Table 9.5 Dematiaceous fungi

Disease	Common organisms
Phaeohyphomycosis	*Exophiala* spp., *Curvularia* spp., *Phialaphora*
Chromoblastomycosis	*Cladophialophora, Fonsecaea pedrosi, Phialaphora verrucosa*
Sporotrichosis	*Sporothrix schenckii*

Table 9.6 Notable dematiaceous fungi

Organism	Yeast/mold	Conidia	Ascospores
Exophiala	Black yeast	Oval shaped, narrow scars, in chains	No
Alternaria	Mold	Long, apical peak, multicellular, transverse, and longitudinal septations. Macroconidia can appear in chains with alternating small end to larger ends (blunt to pointed)	No
Curvularia	Mold	Curved, flat scars, perpendicular septations usually three, with middle section the largest hence making a curve like a boomerang	No
Bipolaris	Mold	Straight, transverse septations can produce germ tubes from both ends	No
Exserohilum	Mold	Boomerang shaped, transverse septations, longer and thinner than bipolaris	No
Pseudallescheria boydii (anamorph *Scedosporium boydii*)	Mold	*Scedosporium boydii* produce oval conidia	Yes, contained in a cleistothecial

Culture from biopsy of lesions may allow for the growth of the mold. Microscopy is required for morphological identification with lactophenol cotton blue preparations. Although pigmented hyphal elements can sometimes be seen in tissue after staining with hematoxylin and eosin (H&E), Fontana–Masson stains are more sensitive. Treatment includes amphotericin B, posaconazole, and voriconazole. Diseases caused by dematiaceous fungi are listed in Table 9.5. Notable dematiaceous fungi are given in Table 9.6.

DERMATOPHYTES

Dermatophytes are fungi which cause infections in keratinized tissues such as hair, skin, and nails. Conidia are the infectious components and will spread by direct contact or via fomites.

Microsporum, *Trichophyton*, and *Epidermophyton*

Disease can spread from host-to-host who may be healthy otherwise. Infection is named based upon location on the person, though different species may cause similar disease in various areas. Skin biopsy at infected site has the best sensitivity for isolating the fungus in culture.

A Wood lamp exam is a diagnostic test that is no longer used extensively that utilizes a black light (emitting wavelength 32–450 nm) to examine skin or hair for *Microsporum* which looks blue-green compared to healthy skin that fluoresces only slight blue. The hair perforation test is used to identify *Trichophyton* species. A blonde hair is placed in water and isolated with the fungus. Hairs are examined with lactophenol blue. *Trichophyton mentagrophytes* will cause pitting and erosion of the hair, while *Trichophyton rubrum* does not.

Microsporum spp.

Microsporum canis causes various fungal infections of the skin including tinea capitis, and tinea corporis. *Microsporum audouinii* can grow in keratinized tissues. Differentiation from *Trichophyton* spp. can be made by the roughness of the macroconidia wall. Macroconidia of *Microsporum canis* are pointed, with an upturned end with hair-like projections on the very end. *M. canis* and *Microsporum gypseum* produce ectothrix invasion of the hair with arthroconidia on the outside of the hair shaft compared to *T. mentagrophytes*.

Trichophyton spp.

T. mentagrophytes complex infection can range from an acute pustular folliculitis to dry lesions. They have the ability to penetrate hair shaft (endothrix). Microconidia appear as grape-like clusters. Spiral-shaped hyphae can occasionally be seen. *Trichophyton tonsurans* causes chronic lesions on the trunk and extremities. *T. tonsurans* causes tinea capitis and invades the hair shaft. *T. rubrum* can cause a chronic type of inflammatory infection as well as tinea corporis, tinea pedis, and onychomycosis. *T. rubrum* is easily spread to other parts of the body and through fomites. The microconidia

Table 9.7 Dermatophyte characteristics

Organism	Microconidia	Macroconidia
Microsporum	Rare, clavate, single	Many, rough, thick-walled, multiseptate, fusiform, spindle shaped
Trichophyton	Many, clavate	Smooth-walled, narrow, lateral, 4–6 cells long
Epidermophyton	None	Smooth-walled, clavate with 1–6 septa

of *T. rubrum* are spread on the hyphae and are often referred to as birds on a wire. The colony has a red pigment.

Epidermophyton floccosum

Epidermophyton floccosum infection can cause skin infection including tinea corporis, tinea curis, tinea pedis, and tinea unguium. Characteristics of dermatophytes are listed in Table 9.7.

KEY POINTS

- The mycotic species include molds, yeasts, and higher fungi. These represent filamentous, spore producing eukaryotic organisms that do not contain chlorophyll.
- Systemic fungal infections are common in individuals who are immunocompromised. Clinical identification through biochemical methods, staining, or molecular techniques is key to successful therapeutic intervention. Microscopy is usually required for identification.
- Eukaryotic unicellular yeasts reproduce by budding, and typically demonstrate hyphal elements. When in spore stage, they are able to survive unfavorable conditions. The *Candida* spp. as well as *Cryptococcus*, represent a common causative opportunistic agents for multiple diseases.
- The thermally dimorphic mold fungi, such as the *Coccidioides* or *Histoplasmosis* spp., have sexual and asexual stages. Transmission can occur through inhalation or contact with environmental organisms. Disease commonly manifests as skin lesions or diffuse erythematous rash, osteomyelitis, or pulmonary infection. Histopathological examination with PAS, GMS, or H&E allow for positive identification from laboratory samples.
- Thermally monomorphic filamentous fungi have sporangiophores which are infectious. These agents cause mucormycoses after inhalation. Effective treatment includes amphotericin B.

- The dematiaceous fungi are melanin producing, and may be acquired from soil or contact with infected plants. Disease is typically subcutaneous in nature, although dissemination may occur in immunocompromised hosts.
- Dermatophytes cause infections in keratinized tissues such as hair, skin, and nails, and are common in otherwise healthy individuals.
- The hyaline molds are environmentally ubiquitous, opportunistic pathogens, which appear as colorless hyaline hyphae when grown on agar medium. The inhaled infectious component is the conidia. The *Aspergillus* spp. are representative, most often causing pulmonary or respiratory disease. Specific stains are useful to identify macroconidia and microconidia.
- Successful treatment for fungi commonly includes use of amphotericin B, fluconazole, itraconazole, posaconazole, or voriconazole.

REFERENCES

[1] Guarner J, Brandt ME. Histopathologic diagnosis of fungal infections in the 21st century. Clin Microbiol Rev 2011;24:247–80.

[2] Naggie S, Perfect JR. Molds: hyalohyphomycosis, phaeohyphomycosis, and zygomycosis. Clin Chest Med 2009;30:337–53 [vii–viii].

[3] Studahl M, Backteman T, Stalhammar F, Chryssanthou E, Petrini B. Bone and joint infection after traumatic implantation of *Scedosporium prolificans* treated with voriconazole and surgery. Acta Paediatr 2003;92:980–2.

[4] Jorgensen JH, Pfaller M, editors. Manual of Clinical Microbiology, vol 1. 11th Edition. Washington DC: ASM Press; 2015.

[5] Larone DH. Medically important fungi: a guide to identification. 5th ed. Washington DC: ASM Press; 2011.

[6] Pfaller MA. Nosocomial candidiasis: emerging species, reservoirs, and modes of transmission. Clin Infect Dis 1996;22(Suppl 2):S89–S94.

[7] Revankar SG, Sutton DA. Melanized fungi in human disease. Clin Microbiol Rev 2010;23:884–928.

CHAPTER 10

Infections Caused by Parasites

Contents

PARASITOLOGY

Introduction

Parasitology is the study of parasites and is traditionally limited to parasitic protozoa, helminths, and arthropods. Human parasitology is focused on

Microbiology and Molecular Diagnosis in Pathology.
DOI: http://dx.doi.org/10.1016/B978-0-12-805351-5.00010-7

medical parasites and includes their morphology, life cycle, and the relationship with host and environment.

Broadly speaking there are two types of parasites: endoparasites and ectoparasites. Endoparasites live within the host. They may be obligate parasites (dependent on their hosts and cannot live without the host), facultative parasites, and accidental parasites. Ectoparasites are parasites which live on the outer surface of the host.

Life cycle of parasites varies. Some have simple life cycle where all the developmental stages are completed in a single host. In others two different hosts are required. One is the definitive host and the other intermediate host. The definitive host is the one which harbors the adult parasite and where the parasite reproduces sexually. The intermediate host is the host which harbors the larval stage or the asexual forms of the parasite. Few parasites require two different intermediate hosts in addition to a definitive host.

Parasites are transmitted by various routes. These include, oral route, penetration by skin or mucous membrane, inoculation by arthropod vectors, and lastly by sexual contact.

Clinical features as a result of parasitic infestations may vary; some are acute, whereas most are chronic.

Diagnosis of parasitic infections depends on clinical diagnosis and laboratory diagnosis. Laboratory diagnosis includes documentation of characteristic forms of the parasites in the feces, urine, sputum, body secretions, or blood. Serologic tests are also available for certain parasites.

PROTOZOA

Protozoa are single-celled organism with a cell membrane, nucleus, and cytoplasm. In some the cytoplasm is divided into ectoplasm and endoplasm. The cytoplasm has various organelles such as mitochondria, endoplasmic reticulum, and Golgi bodies. Protozoa may have locomotory organelles and these include flagella, cilia, and pseudopodia. The nucleus of protozoa contains a karyosome and chromatin. The karyosome may be centrally placed within the nucleus or at the periphery. Most protozoa have a single nucleus. However, ciliates have two nuclei (a micronucleus and a macronucleus). In certain protozoa such as trypanosomes, a nonnuclear DNA containing body is present. This is referred to as a kinetoplast.

Protozoa exists in two stages: Trophozoite form and cyst stage. The trophozoite stage is the reproductive stage and is the state associated with

Table 10.1 Examples of protozoa and their various locations

Parasite	Location
Entamoeba histolytica, Entamoeba coli	Gastrointestinal tract (GT)
Iodamoeba butschlii	GI
Endolimax nana	GI
Blastocystis hominis	GI
Giardia lamblia	GI
Isospora sp.	GI
Blastocystis hominis	GI
Cryptosporidium sp.	GI
Acanthamoeba sp.	Brain
Naegleria fowleri	Brain
Plasmodium vivax	Red cells
Plasmodium malariae	Red cells
Plasmodium falciparum	Red cells
Plasmodium ovale	Red cells
Trichomonas vaginalis	Urogenital tract
Trypanosoma	Blood
Leishmania	Reticuloendothelial system
Toxoplasma gondii	Karyocyte (any nucleated cell)

disease states. The cyst stage is the resting stage and typically replication does not occur at this stage.

Protozoa can be involved in person to person transfer. Examples of this include intestinal amoebae, ciliates, and flagellates. Some protozoa require a definitive host and an intermediate host to complete their life cycle. Some require an insect vector (e.g., *Plasmodium*). Examples of protozoa and their various host locations are summarized in Table 10.1.

Intestinal Protozoa

These may be categorized as:

- Nonflagellated intestinal protozoa.
- Flagellated intestinal protozoa.
- Coccidia.
- Microsporidia.

Nonflagellated intestinal protozoa can be further subdivided into following groups:

- *Entamoeba histolytica.*
- *Entamoeba coli.*
- *Endolimax nana.*

- *Iodamoeba butschlii.*
- *Blastocystis hominis.*

E. histolytica is pathogenic. Other Entamoeba is typically not pathogenic, but may cause diagnostic confusion with *E. histolytica*.

E. histolytica exists as trophozoites or cysts. Morphologic characteristics of the trophozoite stage include:

- 20–30 μm in size.
- Outer clear ectoplasm and inner granular endoplasm.
- A nucleus with a karyosome at the center.
- Clear area around the karyosome.
- May have red cells in the cytoplasm.

Morphologic features of the cyst include:

- Cysts with one to four nuclei, each with central karyosome.
- Early cysts have glycogen vacuoles and chromatoidal bodies.
- Mature cysts have four nuclei.

Humans are infected with food and water contaminated with mature (quadrinucleate) cysts. Infection may also be transmitted by anal-genital and oral-genital route. The cysts excyst in the small intestine thus producing trophozoites. These colonize the mucosal surfaces of the large intestine. In the intestine they may remain asymptomatic. In others they may cause invasive intestinal amebiasis, resulting in blood and mucus in stool with diarrhea (acute amebic dysentery). Also, some intestinal amebiasis may take a chronic form with intermittent diarrhea and abdominal pain. In some the trophozoites may enter into the portal circulation and cause hepatic amebiasis and subsequent amebic liver abscess. Amebic abscess may rupture through the right diaphragm and cause pulmonary amebiasis. Various intestinal nonflagellated protozoa are listed in Table 10.2.

Diagnosis

Various diagnostic methodologies exist but it is best accomplished by the combination of serology or antigen testing together with identification of the parasite.

Stool microscopy—presence of cysts or trophozoites in the stool suggests intestinal amebiasis, but microscopy cannot differentiate between *E. histolytica* and other strains. Organism excretion can vary; a minimum of three specimens on separate days should be sent to detect 85%–95% of infections. Specimens can be concentrated and stained with iodine to detect cysts. To look for trophozoites, a saline wet mount and a fresh smear stained with iron hematoxylin may be performed.

Table 10.2 Intestinal nonflagellated protozoa

Protozoa		Cyst	Trophozoite
Entamoeba histolytica	Pathogenic	10–20 μm, peripheral, fine evenly distributed chromatin; 2–4 nuclei with small central karyosome	12–60 μm, may have red blood cells within cytoplasm
Entamoeba coli	Nonpathogenic, larger than *E. histolytica*	10–35 μm, coarse, clumped unevenly arranged peripheral chromatin, 8–16 nuclei with large eccentric karyosome	15–50 μm, sluggish compared to *E. histolytica*
Endolimax nana	Nonpathogenic	5–10 μm with 2–4 nuclei; no peripheral chromatin	6–12 μm; single nucleus with large central karyosome
Iodamoeba butschlii	Nonpathogenic	5–20 μm; single nucleus with no peripheral chromatin	8–20 μm
Blastocystis hominis	Debatable whether they are intestinal commensals or true pathogens; possible link between this protozoa and irritable bowel syndrome (IBS)		

Stool specimens are frequently positive for blood in the setting of invasive intestinal amebic disease. The presence of ingested erythrocytes is not pathognomonic for *E. histolytica* infection.

Antigen testing—Antigen detection is sensitive, specific, rapid, easy to perform, and can distinguish between *E. histolytica* and *Entamoeba dispar.*

Stool and serum antigen detection assays are commercially available for diagnosis of *E. histolytica* infection. Antigen detection kits using enzyme-linked immunosorbent assay (ELISA), radioimmunoassay, or immunofluorescence have been developed.

Serology—Serology is a useful diagnostic tool for amebiasis. *E. histolytica* infection results in the development of antibodies; *E. dispar* infection does not. Antibodies are detectable within 5–7 days of acute infection and may persist for years. Approximately 10%–35% of uninfected individuals in endemic areas have antiamebic antibodies due to previous infection with *E. histolytica*. Therefore, negative serology is helpful for exclusion of disease, but positive serology cannot distinguish between acute and previous infection.

Molecular methods—Detection of parasitic DNA or RNA in feces via probes can also be used to diagnose amebic infection and to differentiate between the three different strains. Polymerase chain reaction (PCR) techniques can detect *E. histolytica* in stool specimen.

A number of investigators have developed PCR methods for the diagnosis of intestinal amebiasis and differentiation between pathogenic and nonpathogenic amebae although they are not readily available in clinical laboratories [1].

Visual inspection of the colon—Sigmoidoscopy and/or colonoscopy can be performed either to make the diagnosis of amebiasis or to exclude other causes of the patients' symptoms. However, colonoscopy is not recommended as a routine diagnostic approach since intestinal amebic ulcerations increase the likelihood of perforation during instillation of air to expand the colon.

Scrapings or biopsy specimens, best taken from the edge of ulcers, may be positive for cysts or trophozoites on microscopy, and antigen testing for *E. histolytica* may be positive. Colonic lesions in amebic dysentery range from nonspecific mucosal thickening and inflammation to classic flask-shaped amebic ulcers.

Intestinal Flagellated Protozoa

Included in this category are:

- *Giardia lamblia.*
- *Dientamoeba fragilis.*
- *Trichomonas.*
- *Chilomastix.*
- *Balantidium coli.*

Giardia lamblia

Giardia lamblia is a flagellated protozoon that resides in the small intestine causing diarrhea in man. The organism is harbored by many rodents and beavers. Campers frequently develop infection after drinking water contaminated with cysts of *G. lamblia*. Cysts convert to the trophozoite form in the intestine and it resides in the crypts of the duodenum. These trophozoites multiply asexually by binary fission. The trophozoites attach to the duodenal surface, without invasion. The trophozoites are converted to the cyst stage in the lower small intestine. They are passed out in the feces. The cysts can then infect other humans.

Typical morphology of the trophozoites:

- Pear shaped with broad anterior end and tapering posterior end.
- Dorsal convex surface.
- Ventral concave surface, with a sucking disk.
- Bilaterally symmetrical with two nuclei, two axostyles, and four pairs of flagella.

Diagnosis:

- Identification of trophozoites or cysts in feces, duodenal aspirates, or biopsy.
- Direct fluorescent antibody (DFA) or enzyme immunoassay (EIA) for Giardia antigen.

Coccidia and Microsporidia

Included in the coccidia group are *Cryptosporidium*, *Isospora*, and *Cyclospora*.

Cryptosporidium, *Cystoisospora*, *Cyclospora*, and microsporidia infections are seen in HIV/AIDS patients. Franzen and Muller reviewed cryptosporidia and microsporidia waterborne diseases in immunocompromised patients [2].

Cryptosporidium parvum is the only species known to infect man. It infects the small intestine causing diarrhea. Infection is seen mainly in HIV/AIDS patients or those who are immunocompromised. Human is infected with food or drink contaminated with feces which contain oocysts of Cryptosporidium. These oocysts are thick walled and contain four bow shaped sporozoites. In the intestine the sporozoites are released which invade the epithelial lining of the intestine. Within the epithelial cells the sporozoites differentiate to trophozoites. From the trophozoites are derived the merozoites. Some of the merozoites are converted to macro (female) and microgametocytes. Fertilization of the macrogametocytes with microgametocytes results in oocysts. Each oocyst has four sporozoites. If the oocyst is thin walled the sporozoites are released within the intestine and autoinfection ensues. If the oocysts are thick walled they are excreted in the feces.

Diagnosis:

- Direct microscopy of feces: Wet smear, acid fast staining, and direct fluorescent antibody examination.
- Histology of GI tissue.
- Antigen detection.

Cystoisospora

Cystoisospora belli, previously known as *Isospora belli*, may cause watery diarrhea. In general protozoan infection does not cause eosinophilia. A well noted exception to this is infection with *Cystoisospora*. Oocysts are found in the stool of infected individuals, which stain with acid fast stains. The oocysts are large (20–35 μm) and contain one sporozoite.

Naegleria and *Acanthamoeba*

Naegleria and *Acanthamoeba* are free living ameba and are ubiquitous in nature, found mainly in soil and water. *Naegleria fowleri* is found in lakes, swimming pools, tap water, and air conditioning units. *N. fowleri* causes an acute, fulminant, and rapidly fatal meningoencephalitis.

Diagnosis of *N. fowleri*:

- Cerebrospinal fluid (CSF) wet mounts can be stained with Wright Giemsa. Trophozoites are 7–20 μm and resemble white blood cells (WBCs).
- *Naegleria* can be cultured on a bed of *E. coli*.

In healthy individuals certain species of *Acanthamoeba* may result in keratitis. In the immunocompromised certain species can cause granulomatous encephalitis.

The trophozoite of *N. fowleri* typically gains access through the olfactory neuroepithelium when an individual is swimming or diving in water. The trophozoite is typically 10–15 μm in diameter with a prominent nucleus and a large karyosome surrounded by a halo. The trophozoites are actively motile. The trophozoites may be identified in the CSF with a wet mount preparation. PCR studies for detection of parasite DNA is available from Center for Disease Control and Prevention (CDC).

Cyclospora

Cyclospora cayetanensis are found in contaminated food or water. The infected host will shed the noninfective form, the oocyst, in the stool and will sporulate to the infective form only under favorable conditions. Outbreaks have been seen in fresh produce including raspberries and basil. Unsporulated oocysts measure 7–10 μm in diameter. Oocysts can be identified with the dye safranin or a modified acid fast stain.

Microsporidium

Microsporidia are obligate intracellular fungi, but have been historically treated as protozoa. Microsporidia are found in the environment and the entire group contains more than 1200 species from more than 100 genera. Microsporidia contain resistant spores of various sizes as well as a unique polar tubule. The spores range from 1 to 4 μm which is useful for diagnosis. Ingestion of the spores from food and water can cause opportunistic disease. Examples of Microsporidian known to cause disease include *Encephalitozoon intestinalis* causing infection of the GI tract with diarrhea and possible dissemination to ocular, genitourinary, and respiratory tracts, while *Microsporidium ceylonensis* can cause ocular infection. Staining fecal samples to identify Microsporidium species is possible and stains the spore wall but cannot differentiate among the microsporidia to the species level. Immunofluorescence assays are available and the CDC offers a PCR that can differentiate some of the microsporidia.

Leishmania

Leishmaniasis is a disease that is transmitted by sandflies. *Leishmania donovani* causes visceral leishmaniasis also known as kala-azar. In the sand fly, the promastigote are found in the GI tract. The promastigote form is spindle shaped with a single flagellum. The promastigote form is injected into humans when a sand fly bites. The promastigotes are taken up by the reticuloendothelial system and become transformed into the amastigote form. Amastigote forms are also found in endothelial cells and polymorphonuclear neutrophils (PMNs). The amastigote form has a nucleus and a kinetoplast at right angle to the nucleus. The kinetoplast is a slender, rod-shaped structure. The amastigote form is devoid of flagella. The amastigote form multiplies in the human and results in visceral leishmaniasis, characterized by splenomegaly, hepatomegaly, bone marrow involvement, and lymphadenopathy. Other forms of leishmaniasis are:

- Cutaneous leishmaniasis: Cutaneous lesions where the sand fly has injected the promastigotes.
- Postkala-azar dermal leishmaniasis: Cutaneous lesions seen after completion of treatment of visceral leishmaniasis.

Diagnosis:

- Direct microscopy: Direct microscopy of peripheral blood (buffy coat), bone marrow aspiration, splenic aspiration, liver, or lymph node biopsy. It is important to note that *Leishmania* may appear similar to *Histoplasma*, but have a kinetoplast, a network of circular DNA found in the mitochondria.
- Skin biopsy.
- Serology: For example, ELISA, not readily available.
- Leishmanin skin test.

Schriefer et al. reviewed recent developments in diagnostic and therapeutic approach to human leishmaniasis [3].

Trypanosomes

Trypanosomes are flagellated protozoa. *Trypanosoma cruzi* causes Chagas disease (a.k.a. South American trypanosomiasis) and is transmitted by reduvid bugs.

Trypanosoma brucei causes sleeping sickness (a.k.a. African trypanosomiasis) and is transmitted by the Tsetse fly. There are two subspecies of *T. brucei*: *T.b. gambiense* which causes West African sleeping sickness and *T.b. rhodesiense* which causes East African sleeping sickness.

Chagas Disease

Humans are infected when a reduvid bug bites and releases trypomastigotes in its feces near the site of wound. Trypomastigotes enter the body through the wound. Trypomastigote is the flagellated form of the parasite. The trypomastigote form is typically C shaped. It has a flagellum at the anterior end and the flagella traverses on the surface of the parasite as an undulating membrane. There is a central nucleus and a kinetoplast at the posterior end. The posterior end is wedge shaped. The trypomastigote form may be found in the blood of humans.

Inside the host the trypomastigote invades cells and is converted to amastigotes. Amastigotes are nonflagellated organisms with a nucleus and kinetoplast. They are morphologically similar to *Leishmania* species. The amastigotes multiply asexually and can be differentiated into trypomastigotes. These can be released into the circulation where they will infect new cells. Reduvid bugs will be infected when feeding on human blood with circulating trypomastigotes.

Clinical features of Chagas disease are:

- Localized swelling at the site of inoculation (Chagoma).
- When the parasite is inoculated in the conjunctiva, there occurs swelling of the periocular tissue (Romaña's sign).
- Cardiomyopathy.
- Megaesophagus and megacolon.

Diagnosis:

- Microscopy of blood.

Sleeping Sickness

Humans are infected when a *Tsetse* fly (either male or female) bites. The trypomastigotes that are transmitted multiply invade various organs including, heart, connective tissue and central nervous system. The morphology of the trypomastigote is similar to that of the trypomastigote form of *T. cruzi*. However, the trypomastigote form of *T. brucei* is slender and fusiform. They have a flagellum at the anterior end with an undulating membrane. There is a central nucleus and a kinetoplast at the posterior end.

Clinical features are:

- Chancre: At the site of inoculation, a painful, erythematous swelling.
- Lymphadenopathy.
- Hepatosplenomegaly.
- Meningoencephalitis: Somnolence, coma, and death.

Diagnosis:

- Microscopy of chancre fluid, lymph node aspirates, blood, and bone marrow.

Trichomonas vaginalis

Trichomonas vaginalis is another flagellated protozoan and causes the sexually transmitted disease, trichomoniasis.

T. vaginalis exists in the lower genital tract of the female and the urethra and prostate in the male. Only the trophozoite forms exist. There is no cyst stage of this protozoon. Transmission is by sexual intercourse.

Morphology of *T. vaginalis* protozoa:

- Pear shaped and motile.
- Single nucleus.
- Four anterior flagella.
- Single lateral flagellum which forms an undulating membrane.
- Central axostyle.

Diagnosis:

- Microscopic examination for trophozoites from genital specimens wet prep visualize motile organisms.
- Culture.
- Molecular detection female genital specimens.

Plasmodium

Five species of *Plasmodium* infects humans. These are *Plasmodium vivax*, *Plasmodium ovale*, *Plasmodium malariae*, *Plasmodium knowlesi*, and *Plasmodium falciparum*. Worldwide the greatest mortality is due to *P. falciparum*. *P. knowlesi* and *P. vivax* also cause significant mortality and morbidity.

For these *Plasmodium* species female *Anopheles* mosquitoes are the definitive host. Sexual reproduction takes place in the mosquitoes. Humans are infected with a bite of the mosquito. The mosquito transmits the sporozoite form of the parasite into humans. These are spindle shaped with a single nucleus. The malarial parasites undergo an exo-erythrocytic cycle where they infect hepatocytes. Within the hepatocytes these sporozoites undergo cell division and result in numerous merozoites within a schizont. The hepatocytes rupture and these merozoites are released into the circulation and these then infect red blood cells (RBCs). In cases of *P. vivax* and *P. ovale* the sporozoites may remain dormant within hepatocytes; these are referred to as hypnozoites. Hypnozoites remain in the liver, only to be converted to merozoites at a later stage.

In the RBCs (erythrocytic cycle), the merozoites are transformed into trophozoites. The trophozoites ingest host cytoplasm which forms a food vacuole in the trophozoite. This gives the appearance of a ring form with the nucleus at one edge. With time the food vacuole becomes smaller and hemozoin pigment granules appear. Hemozoin pigments are derived from hemoglobin. Within the red cells the trophozoites multiply to form schizonts. The red cells then rupture to release merozoites which can infect new red cells. Some merozoites instead of undergoing erythrocytic schizogony become transformed to macro (female) and micro (male) gametocytes. These gametocytes when ingested by a mosquito completes the sexual part of the life cycle within the mosquito.

Clinical Features of Malaria

There is a period known as the incubation period between sporozoites being introduced to humans by the mosquito bite to onset of clinical features. In falciparum malaria, the incubation period is about a week. In nonfalciparum malaria, it varies from 9 to 14 days.

Clinical features include:

- Fever, preceded by chills. In nonimmune patients, fever may occur daily. In semiimmune patients, there may be febrile episodes (fever every 48 hours a.k.a. tertian for cases of *P. falciparum*, *P. vivax*, and *P. ovale* and fever every 72 hours a.k.a. quartan for *P. malariae*). Periodic fever is not seen in *P. knowlesi*.
- Headache.
- Malaise.
- Loss of appetite.
- Anemia.

Hypnozoites of *P. vivax* and *P. ovale* in the liver can at a much later time become active and multiply and be released from hepatocytes to infect red cells. This results in relapse of infection with these two species. Thus treatment should include eradication of hypnozoites from the liver.

P. malariae can remain in the blood of humans for years only to become active at a later stage. This is referred to as recrudescence.

P. falciparum does not exhibit relapse or recrudescence.

Cerebral malaria is a serious condition seen with *P. falciparum* infection. Microvessels within the brain become occluded with infected red cells, resulting in coma and death. Grau and Craig reviewed pathogenesis of cerebral malaria [4].

Blackwater fever is also seen with *P. falciparum*. Here there is significant hemolysis and release of hemoglobin into the circulation. Free hemoglobin and its breakdown product color the urine dark. Free hemoglobin also results in acute tubular necrosis.

Tropical splenomegaly syndrome (hyperreactive splenomegaly) is seen in cases with chronic malaria. There is overproduction of IgM due to repeated infection, immune complex formation, and stimulation of the splenic reticuloendothelial cells, resulting in splenomegaly. Parasites are not seen in the blood.

Diagnosis of Malaria

- *Direct microscopy*: Thick and thin films made from peripheral blood. Thick films are ideal to detect presence of malarial parasites. Thin films are ideal for species identification as well estimation of parasite load. The CDC recommends a thick and thin smear every 12–24 hours for a total of three sets, before ruling out malaria. Morphologically

P. malariae and *P. knowlesi* are similar. PCR and gene sequencing will be able to distinguish the two species.

- *Detection of malarial antigen*: Detection of malarial antigens forms the basis of rapid detection tests. The antigens which may be targeted include histidine rich protein 2 (HRP2), plasmodium lactate dehydrogenase, and aldolase. HRP2 antigens are only seen with *P. falciparum*.
- Serology.
- PCR studies.

Morphologic diagnostic features of *P. falciparum*:

- Red cells not enlarged.
- Rings appear fine/delicate, several per cell.
- Some rings with two chromatin dots.
- Presence of marginal/applique forms.
- Unusual to see developing forms.
- Crescent-shaped gametocytes.
- Maurer's dots may be present.

Morphologic diagnostic features of *P. vivax*:

- Red cells containing parasites usually enlarged.
- Schüffner's dots (bright red dots) frequently present.
- Mature ring forms large and coarse.
- Developing forms frequently present.

Morphologic diagnostic features of *P. malariae*:

- Ring forms have squarish appearance.
- Band forms characteristic.
- Mature schizonts may have typical "daisy head" appearance with <10 merozoites.
- Red cells not enlarged.
- Chromatin dot may be on inner surface of the ring.

Morphologic diagnostic features of *P. ovale*:

- Red cells enlarged.
- Comet forms common.
- Rings large and coarse.
- Schüffner's dots prominent when present.
- Mature schizonts similar to *P. malariae*, but larger and more course. The Duffy antigen is a glycoprotein on the surface of red cells. The antigen acts as a receptor for *P. vivax*. Thus, RBCs that lack Duffy antigens are relatively resistant to infection by *P. vivax*. Sickle cell trait patients also have protective advantage against malaria. Thus, there is increased frequency of sickle cell trait individuals in malaria endemic areas.

Toxoplasma gondii

Felines are the only animals in which *Toxoplasma gondii* is able to complete its reproductive cycle. *T. gondii* in felines, once ingested infects GI tract epithelium. Oocysts are passed out in the feces. The oocysts are able to infect humans. From the GI tract widespread dissemination throughout the body takes place. The oocysts will be transformed into the trophozoite form which multiplies within the host cell. Once the host cell ruptures the trophozoites are released to infect new cells. At times, these trophozoites may encyst and these forms may remain dormant in the body. Common sites of encystations are brain, cardiac muscle, and skeletal muscle.

T. gondii infection may occur by:

- Ingestion of oocysts originating from cat feces.
- Ingestion of cysts in meat.
- Vertical transmission (transplacental transfer of trophozoites).
- Blood transfusion or organ transplantation from infected donors.

Clinical Features

In the immunocompetent individual, most cases of infection of *T. gondii* is asymptomatic. In a few, there is usually bilateral symmetrical cervical lymphadenopathy. In immunocompromised, encephalitis is a serious feature. *T. gondii* is one of the most common causes of chorioretinitis in immunocompetent individuals. Infection typically occurs congenitally or postnatally.

Congenital Toxoplasmosis

This is due to acute primary infection by the mother during pregnancy. Abortion, intrauterine death, neurologic, and ocular pathology dominates the clinical picture. These features are most often seen with infection during the first trimester of pregnancy.

Diagnosis:

- Clinical/radiologic (characteristic ring enhancing lesion).
- Serology (e.g., ELISA) IgG, IgM, and IgA.
- Histology.
- Blood test using PCR.

TREMATODES

Trematodes are also known as flukes. In humans flukes may be found in a variety of organs including the intestine, lungs, and liver. Trematodes are flat and leaf like with bilaterally symmetrical body. They are all hermaphrodites except *Schistosoma* species.

The definitive host is man where sexual reproduction takes place. For trematodes the number of intermediate hosts may be more than one. Life cycles of trematodes are summarized in Table 10.3.

The trematodes of importance for human infestation are:

- *Clonorchis sinensis.*
- *Fasciola hepatica.*
- *Fasciolopsis buski.*
- *Paragonimus westermani.*
- *Schistosoma japonicum.*
- *Schistosoma hematobium.*
- *Schistosoma mansoni.*

Diagnosis of trematodes is primarily by morphologic identification of eggs either in the stool (*C. sinensis*, *F. hepatica*, *F. buski*, *S. mansoni*, and *S. japonicum*), sputum (for *P. westermani*) and urine for *S. hematobium*. Morphological features of eggs of trematodes are summarized in Table 10.4.

CESTODES

Cestodes are long segmented tape-like worms. The adult worms are flat and white in color. The adult worm has a scolex, neck, and strobilus. The scolex contains organs which facilitate attachment to the host tissue. Examples of this include suckers or hooks. The neck contains germinal cells which can give rise to new proglottids. Proglottids are individual segments of the strobilus and contain reproductive organs of both sexes. The proglottids at the posterior end are mature in comparison to the ones at the anterior end. Cestodes are hermaphrodites with a few exceptions. Cestodes complete their life cycle typically in two or three hosts. An exception is *Hymenolepis nana* which needs a single host only. Life cycles of cestodes are summarized in Table 10.5.

We will consider the following cestodes:

- *Diphyllobothrium latum.*
- *Taenia saginata.*
- *Taenia solium.*
- *H. nana.*
- *Dipylidium caninum.*
- *Echinococcus granulosus.*

Table 10.3 Life cycle of trematodes

Trematode	Human infected by	Location of adult worm in human	Definitive host	First intermediate host	Second intermediate host	Comments
Clonorchis sinensis also known as Chinese liver fluke	Metacercariae in undercooked fish	Biliary tree	Human; Passes eggs in feces. Eggs excyst to form miracidium which infects snails	Snail; which releases cercariae	Fish; infected by cercariae. Metacercariae form in flesh of fish	Associated with cholangiocarcinoma
Fasciola buski/ *Fasciola hepatica*	Metacercariae in underwater plants	Small intestine (*F. buski*) and biliary tree (*F. hepatica*)	Human and pigs; passes eggs in feces. Eggs excyst to form miracidium which infects snails	Snail; which releases cercariae	None. Metacercariae develop from cercariae	
Paragonimus westermani	Metacercariae in crustacean	Lungs	Human; Passes eggs in feces. Eggs excyst to form miracidium which infects snails	Snail; which releases cercariae	Crustaceans; infected by cercariae. Metacercariae form in crustaceans	Other organs such as brain and muscle may also be involved

(*Continued*)

Table 10.3 Life cycle of trematodes (Continued)

Trematode	Human infected by	Location of adult worm in human	Definitive host	First intermediate host	Second intermediate host	Comments
Schistosoma hematobium	Penetration of cercariae through intact skin	Blood vessels of bladder, prostate, and uterine plexus	Human; passes eggs in urine. Eggs excyst to form miracidium which infects snails	Snail; which releases cercariae	None	
Schistosoma mansoni	Penetration of cercariae through intact skin	Mesenteric veins	Human; passes eggs in stool. Eggs excyst to form miracidium which infects snails	Snail; which releases cercariae	None	
Schistosoma japonicum	Penetration of cercariae through intact skin	Mesenteric veins	Human; passes eggs in stool. Eggs excyst to form miracidium which infects snails	Snail; which releases cercariae	None	

Table 10.4 Morphology of eggs of trematodes

Chlonorchis	*Fasciola*	*Paragonimus*	*Schistosoma hematobium*	*Schistosoma japonicum*	*Schistosoma mansoni*
25–35 μm, operculated egg; bile stained. A small knob at the opposite end of the operculum	130–150 μm; operculated egg	80–120 μm; operculated egg	110–170 μm; egg with terminal spine	70–100 μm; egg without spine	110–180 μm; egg with lateral spine

Table 10.5 Life cycle of cestodes

Cestode	Humans infected by	Location of adult worm	Definitive host	First intermediate host	Second intermediate host	Comments
Diphyllobothrium latum also known as fish tapeworm	Raw or undercooked fish which has plerocercoid larva	Intestine	Human: passes eggs or proglottids in stool	Coracidia derived from eggs ingested by crustacean	Procercoid larva develops from coracidia; crustaceans with procercoid larva ingested by fish where plerocercoid larva develops	Scolex has two suckers
Taenia saginata also known as beef tapeworm	Raw or poorly cooked beef with *T. saginata* larva	Intestine	Human: passes eggs or proglottids in stool	Cattle; oncospheres derived from eggs invade intestinal wall and migrate to tissue including muscle		Scolex has four suckers; no hooks

Taenia solium also known as pork tapeworm	Raw or poorly cooked pork with *T. solium* larva (cysticerci)	Intestine	Human: Passes eggs or proglottids in stool	Pigs; oncospheres derived from eggs invade intestinal wall and migrate to tissue including muscle		Scolex has four suckers, rostellum and double row of hooks If eggs are ingested by humans then cysticerci may develop in various human tissues. This is cysticercosis (e.g., neurocysticercosis); not seen with *T. saginata*
Hymenolepis nana	Food or unclean hands with eggs from stool	Intestine	Human; passes egg in stool			Common in children
Dipylidium caninum	Fleas with cysticerci	Intestine	Human, cats, dogs: Passes eggs in stool	Fleas; eggs are ingested by larval stage of fleas; oncospheres derived from eggs invade intestinal wall and develop into cysticerci in body cavity of flea		Children affected

(*Continued*)

Table 10.5 Life cycle of cestodes (Continued)

Cestode	Humans infected by	Location of adult worm	Definitive host	First intermediate host	Second intermediate host	Comments
Echinococcus granulosus also known as dog tapeworm	Eggs from dog stool	Intestine of dogs and canines	Dogs and canines; man is an accidental host			When eggs are ingested by humans, oncospheres are derived from these eggs in the intestine. They penetrate the intestinal wall and reach the liver or lung. The larva derived from the oncospheres is hydatid cysts

Diagnosis of cestodes:

Cestode eggs, mature proglottids, and scoleces are used for diagnosis. A wet mount preparation of stool sample is used.

- The eggs of *Taenia* are indistinguishable.
- The proglottids of *Taenia*, *Diphyllobothrium*, and *Dipylidium* are large (0.5–1.5 cm in length).
- The proglottids of *Hymenolepis* are much smaller, up to 1 mm.
- The proglottids of *T. saginata* have more than 14 ovarian branches.
- The proglottids of *T. solium* have less than 14 ovarian branches.
- The proglottids of *D. latum* have a central rosette.
- The scolex of *Taenia* have suckers.
- The scolex of *T. solium* has hooks.

HYDATID CYST

The typical hydatid cyst has two layers in its wall. It is unilocular and fluid filled. The inner layer is also known as the germinal layer. From the germinal layer protoscolex, brood capsules and daughter cysts may develop. When these structures breakdown they give rise to hydatid sand within the hydatid fluid. Hydatid fluid may give rise to anaphylactic reactions.

NEMATODES

There are thousands of different species of nematodes. However, only handfuls are known to cause human infestation. The ones that do cause human infestation may be broadly divided into three categories:

- Intestinal nematodes.
- Tissue nematodes.
- Nematodes affecting blood and tissues.

The intestinal nematodes affecting humans are:

- *Ascaris lumbricoides* (also known as roundworm).
- *Ancylostoma duodenale* (old world hookworm) and *Necator americanus* (new world hookworm).
- *Enterobius vermicularis* (also known as pinworm).
- *Trichuris trichiura* (also known as whipworm).

The life cycle of *E. vermicularis* and *T. trichiura* is simple and similar. The adult worms typically reside in the large intestine. Eggs are passed in stool. Eggs can then be ingested by humans and subsequently the eggs are hatched in the small intestine to yield larva. These then mature to adult worms which reside in the large intestine. *E. vermicularis* eggs are typically deposited in the perianal fold by the gravid females. This causes perianal itching. The eggs may be picked up by fingernails and can reinfect the same individual. This is referred to as autoinfection.

A. lumbricoides, adult worms reside in the small intestine. Eggs released pass out in stool. Once ingested by humans, larvae hatch out of the eggs in the small intestine. The larvae are capable of penetrating the intestinal wall and enter the circulation. They are carried through the portal circulation to the right heart and then onto the lungs. The larvae move up the respiratory tree and are subsequently swallowed, where they mature to form adult worms residing in the small intestine. Clinical features related to migration through the lungs are referred to as Loeffler's syndrome. The adult worm in the intestine may give rise to abdominal pain, malabsorption, malnutrition, biliary obstruction, and pancreatitis.

A. duodenale and *N. americanus* adult worms reside in the small intestine. Eggs released pass out in stool, hatch, and release larva in the environment. The first larval stage is known as rhabditiform larva which matures into the infective filariform larva. Filariform larva penetrates the skin of humans and gain access into the circulation. They reach the lungs and subsequently the larva move up the respiratory tree and is subsequently swallowed, where they mature to form adult worms residing in the small intestine. The adult worms of *A. duodenale* have a buccal capsule with two pairs of curved teeth whereas *N. americanus* have semilunar cutting plates. Since these are used by the hookworms they result in GI bleeding and chronic blood loss.

Diagnosis of these intestinal nematodes is made by microscopic examination of stool. In the case of *E. vermicularis* a tape is attached to the perianal region to acquire the eggs. Stool may not contain eggs for diagnosis. Morphology of eggs of intestinal nematodes is given in Table 10.6.

Tissue Nematodes

Clinically important tissue nematodes include:

- *Dracunculus medinensis* (also known as guinea worm).
- *Trichinella spiralis*.
- Nematodes causing cutaneous and visceral larva migrans.

Dracunculus medinensis

Humans are infected when they drink water with crustaceans which are infected with larva of *D. medinensis*. The larvae are released in the small

Table 10.6 Morphology of eggs of intestinal nematodes

Intestinal nematode	Eggs
Enterobius vermicularis	Ovoid, asymmetrically flattened on one side; covered by a colorless thick shell
Trichuris trichiura	Barrel shaped with two polar plugs. Yellow brown and double shelled
Ascaris lumbricoides	Eggs are mamillated and thick shelled
Ancylostoma duodenale and *Necator americanus*	Oval, thin shelled and colorless. Eggs contain two to four blastomeres. There is a clear space between developing embryo and shell wall

intestine and penetrate the wall of the intestine to enter the abdominal cavity. Adult worms develop and female adult worms once fertilized migrate to subcutaneous tissue. This is typically in the lower extremities. A papule usually results at the site where the female worms rest. Upon contact with water the female worm discharges larva.

Trichinella spiralis

Pig is the primary host of *T. spiralis*. Humans are accidental hosts and also a dead end host. Humans are infected with undercooked pork meat with encysted larva is skeletal muscle. The larvae are released in the small intestine. Adult worms develop in the small intestine. The gravid female burrows in the intestinal mucosa, discharging larva which enter the circulation and find its way to skeletal muscle where they may remain viable for years. Diagnosis of trichinosis is by muscle biopsy and serology.

Cutaneous Larva Migrans

This is a condition where a larva gains access through the skin of humans but is unable to complete the life cycle. The larva migrates in the skin causing skin eruptions and irritation. The best examples of nematode larvae which cause cutaneous larva migrans are that of *Ancylostoma caninum* and *Ancylostoma braziliense*. These normally infect cats and dogs.

Cutaneous larva migrans can also be caused by cercaria of trematodes, unable to complete their life cycle in humans. Examples are *Schistomas* spp. and *Trichobilharzia*.

Lastly cutaneous larva migrans can also be caused by plerocercoids of *Spirometra mansoni* (a Cestode) penetrating the skin of humans.

Visceral Larva Migrans

Just like cutaneous larva migrans, visceral larva migrans is due to larva which enters the human body, finds their way into various viscera but is unable to complete the life cycle. Larva of certain nematodes, metacercaria of certain trematodes, and plerocercoids of certain cestodes are all implicated in this type of pathology. Relatively well-known examples include *Toxocara canis*, *Toxocara catis*, and *Gnathostoma*.

Blood and Tissue Nematodes

These include:

- *Wuchereria bancrofti*.
- *Brugia malayi*.
- *Onchocerca volvulus*.
- *Loa loa*.
- *Mansonella ozzardi*.

Blood and tissue nematodes are listed in Table 10.7.

Table 10.7 Blood and tissue nematodes

Nematode	Humans infected by	Site of infestation of adult worms	Comments
Wuchereria bancrofti	Mosquitoes: *Anopheles*, *Culex*, and *Aedes*	Lymphatics	Lymphatic obstruction can cause lymphedema and elephantiasis
Brugia malayi	Mosquitoes	Lymphatics	Lymphatic obstruction can cause lymphedema and elephantiasis
Onchocerca volvulus	Black fly	Eyes; microfilariae not found in blood	Also known as river blindness
Loa loa	Deer flies (Chrysops)	Subconjunctival tissue	
Mansonella ozzardi	Black flies and biting midges	Lymphatics, thoracic, and peritoneal cavities	

Table 10.8 Morphology of microfilariae found in the blood of humans

Wuchereria bancrofti	Sheathed microfilariae; tail pointed and clear (devoid of nuclei)
Brugia malayi	Sheathed microfilariae; tail pointed with two nuclei
Loa loa	Sheathed microfilariae; tail blunt with nuclei
Mansonella ozzardi	Unsheathed; tail pointed and clear

Life Cycle of Blood and Tissue Nematodes

Humans are infected by a mosquito bite or fly. The infective form for humans is a third stage larva. Adult worms develop which reside in different areas of the body. The adult worm releases microfilariae and these are picked by the arthropod vectors, where the asexual part of the life cycle is completed.

Diagnosis of the above infections is by visualization of the microfilariae in the blood and/or serology. Morphology of microfilariae found in blood of humans is listed in Table 10.8.

KEY POINTS

- The nucleus of protozoa contains a karyosome and chromatin. The karyosome may be centrally placed within the nucleus or at the periphery. Most protozoa have a single nucleus. However, ciliates have two nuclei (a micronucleus and a macronucleus). In certain protozoa such as trypanosomes, a nonnuclear DNA containing body is present. This is referred to as a kinetoplast.
- *E. histolytica* is pathogenic. Other *Entamoeba* spp. are typically not pathogenic, but may cause diagnostic confusion with *E. histolytica*.
- *G. lamblia* is a flagellated protozoon that resides in the small intestine causing diarrhea in man. The organism is harbored by many rodents and beavers. Campers frequently develop infection after drinking water contaminated with cysts of *G. lamblia*.
- *Cryptosporidium*, *Cystoisospora*, *Cyclospora*, and microsporidia infections are seen in HIV/AIDS patients.
- *Naegleria* and *Acanthamoeba* are free living ameba and are ubiquitous in nature, found mainly in soil and water. *N. fowleri* is found in lakes, swimming pools, tap water, and air conditioning units. *N. fowleri* causes an acute, fulminant, and rapidly fatal meningoencephalitis.
- Leishmaniasis is a disease that is transmitted by sandflies. *L. donovani* causes visceral leishmaniasis also known as kala-azar.

- Trypanosomes are flagellated protozoa. *T. cruzi* causes Chagas disease (a.k.a. South American trypanosomiasis) and is transmitted by reduviid bugs.
- *T. brucei* causes sleeping sickness (a.k.a. African trypanosomiasis) and is transmitted by the *Tsetse* fly. There are two subspecies of *T. brucei*, *T.b. gambiense* which causes West African sleeping sickness and *T.b. rhodesiense* which causes East African sleeping sickness.
- *T. vaginalis* is another flagellated protozoan and causes the sexually transmitted disease, trichomoniasis.
- Five species of *Plasmodium* infects humans. These are *P. vivax*, *P. ovale*, *P. malariae*, *P. knowlesi*, and *P. falciparum*. Worldwide the greatest mortality is due to *P. falciparum*. *P. knowlesi* and *P. vivax* also cause significant mortality and morbidity. For these *Plasmodium* species female *Anopheles* mosquitoes are the definitive host.
- In cases of *P. vivax* and *P. ovale* the sporozoites may remain dormant within hepatocytes; these are referred to as hypnozoites.
- *P. malariae* can remain in the blood of humans for years only to become active at a later stage. This is referred to as recrudescence.
- Cerebral malaria is a serious condition seen with *P. falciparum* infection. Microvessels within the brain become occluded with infected red cells, resulting in coma and death.
- Blackwater fever is also seen with *P. falciparum*. Here there is significant hemolysis and release of hemoglobin into the circulation. Free hemoglobin and its breakdown product color the urine dark. Free hemoglobin also results in acute tubular necrosis.
- Trematodes are also known as flukes. In humans flukes may be found in a variety of organs including the intestine, lungs, and liver. Trematodes are flat and leaf like with bilaterally symmetrical body. They are all hermaphrodites except *Schistosoma* species.
- Diagnosis of trematodes is primarily by morphologic identification of eggs either in the stool (*C. sinensis*, *F. hepatica*, *F. buski*, *S. mansoni*, and S. *japonicum*), sputum (for *P. westermani*), and urine for *S. hematobium*.
- Cestodes are long segmented tape-like worms. The adult worms are flat and white in color. The adult worm has a scolex, neck, and strobilus. The scolex contains organs which facilitate attachment to the host tissue. Examples of this include suckers or hooks. The neck contains germinal cells which can give rise to new proglottids. Proglottids are individual segments of the strobilus and contain reproductive organs of both sexes. The proglottids at the posterior end are matured in

comparison to the ones at the anterior end. Cestodes are hermaphrodites with a few exceptions.

- Eggs, mature proglottids, and scolices are used for diagnosis for cestodes. A wet mount preparation of stool sample is used.
- The larvas derived from the oncospheres of *E. granulosus* are hydatid cysts. The typical hydatid cyst has two layers in its wall. It is unilocular and fluid filled. The inner layer is also known as the germinal layer. From the germinal layer protoscolex, brood capsules and daughter cysts may develop. When these structures breakdown they give rise to hydatid sand within the hydatid fluid. Hydatid fluid may give rise to anaphylactic reactions.
- Diagnosis of intestinal nematodes are microscopy of stool.
- Cutaneous larva migrans is a condition where a larva gains access through the skin of humans but is unable to complete the life cycle. The larva migrates in the skin causing skin eruptions and irritation.
- Visceral larva migrans is due to larva which enters the human body finds their way into various viscera but is unable to complete the life cycle. Larva of certain nematodes, metacercaria of certain trematodes and plerocercoids of certain cestodes are all implicated in this type of pathology.
- Diagnosis of blood and tissue nematodes is typically done by identification of microfilariae.

REFERENCES

[1] van Lieshout L, Verweij JJ. Newer diagnostic approaches to intestinal protozoa. Curr Opin Infect Dis 2010;23:488–93.
[2] Franzen C, Muller A. Cryptosporidia and microsporidia—waterborne diseases in the immunocompromised patients. Diagn Microbiol Infect Dis 1999;34:245–62.
[3] Schriefer A, Wilson ME, Carvalho EM. Recent developments leading towards a paradigm switch in the diagnosis and therapeutic approach to human leishmaniasis. Curr Opin Infect Dis 2008;21:483–8.
[4] Grau GE, Craig AG. Cerebral malaria pathogenesis: revisiting parasite and host contributions. Future Microbiol 2012;7:291–302.

CHAPTER 11
Antigen and Antibody Testing

Contents

INTRODUCTION

Tests for detection of antibodies, antigens, and immunologic evidence of microbial infections are reviewed in this chapter. Appropriate uses of these tests for diagnosis of infectious diseases are also discussed. Immune assays and experimental systems are designed to assist in assessment of pathogenic response, and are instrumental in our understanding of the pathophysiology of disease. The exquisite specificity of the antibody allows its use as an efficient analytical tool in biomedical research and diagnostic investigation.

The heart of the immune assays is the ability of antibodies to specifically recognize target shapes (antigens) unique to pathogenic molecules. The interactions between antigen and antibody form the basis of many quantitative and qualitative diagnostic assays which involve noncovalent binding of antigenic determinants (epitopes) to the variable regions of both the heavy and light immunoglobulin chains. The use of antibodies

Microbiology and Molecular Diagnosis in Pathology.
DOI: http://dx.doi.org/10.1016/B978-0-12-805351-5.00011-9

has expanded from simple diagnostics to elucidation of gene function, localization of gene products, and rapid screening of biological effectors for drug discovery and testing.

The antibody titer reflects the measured amount of antibodies in the blood, described as a relative dilution factor against a particular type of antigen (pathogen, molecule, or substance). Antibodies directed against epitopes on multideterminant antigens lead to agglutination reactions. The most widely known agglutination reaction is Coombs' test, originally used to detect autoantibodies against host red blood cells (RBCs). Many diseases and drugs can lead to production of these antibodies. In direct agglutination tests the agglutinating antibodies react with antigens on the surface of bacterial cells (e.g., Brucella and Tularemia febrile agglutinin tests), or even erythrocytes (direct hemagglutination test) to form visible clumps of particles. In passive agglutination tests, agglutinating antibodies react with antigens coupled to carrier particles such as gelatinous beads to form visible aggregates.

Antibodies in serum are by nature polyclonal, with potential to recognize more than 10^{15} different antigenic epitopes. Exposure to foreign molecules (nonself), either by infection or by immunization, will increase the relative balance and titer of polyclonal antibodies toward a specific antigen. In 1975, Köhler and Milstein described a procedure for creating fused cells that secrete antibodies of a single defined specificity [1]; in essence describing the first monoclonal antibody. Monoclonal antibodies are well suited as diagnostic reagents for sensitive and specific detection of defined molecular epitopes.

GENERAL TECHNIQUES FOR ANTIGEN RECOGNITION

Immunochemical-based techniques offer advantages in that they allow specific antigen recognition for on-site detection within the laboratory setting, allowing sensitive and selective determination of target factors present in, or released from, potential pathogens. Below we review some of the technology underlying main antigen assays currently in laboratory use.

Antibody–Antigen Precipitation Reactions

Often, precipitation reactions are used for analysis in situations where only qualitative detection is required. Reactions can be performed in a gel matrix to limit the rate of diffusion of reactants and hold the precipitate so that it is effectively immobilized for visualization. Several methods are widely used in medicine for analysis of hormones, enzymes, toxins, and for analysis of immune system products.

Latex Agglutination

The latex agglutination reaction is a variation on the clumping ability observed when a sample containing the specific antigen is mixed with an antibody coated on the surface of latex particles. The agglutination reaction may also be reversed, with specific antigen coated on the latex particle, and testing done on a sample to determine the presence of specific antibodies. The use of synthetic beads offers the advantages of consistency, uniformity, and stability. Furthermore, the latex agglutination assay offers a rapid advantage for quick results, often determined within minutes. Other advantages include versatility of synthetic beads for use (allowing binding of chemically complex antigens), simplicity of design, ease of performance, and ability to work using small quantities of sample. Finally, this type of reaction does not by definition require expensive equipment for assessment of samples. A word of caution should be noted in that contaminations in sample preparations may lead to interference and false negatives.

Lateral Flow

The lateral flow assay (LFA) is a paper-based platform for the detection and quantification of molecules in complex mixtures [2]. It is relatively rapid, and results may be obtained within a 30 minutes time period. A microstructured polymer is used to transport fluid, acting first as a sponge to hold sample (such as urine or serum). Fluid then migrates to secondary conjugate pad in which a dried bio-active particle resides in a salt–sugar matrix, thus permitting optimized chemical reactivity between the target antigen and its immobilized partner antibody. The flowing fluid dissolves the matrix salts while also allowing transport through the pored structure. Antigens bind to particles as they migrate with the fluid flow through reactive zones. A third level of capture, usually onto an immobilized strip, permits binding of the complex. The LFA can operate as a capture assay where the antigen is sandwiched between antibodies directed toward multiple determinants on the molecule targeted for assessment. Alternatively, the assay can be set as a competitive assay, quantitating antigen presence through competitive assessment.

Rapid Chromatographic Immunoassay

The rapid chromatographic immunoassay is extremely useful for rapid assessment of mycotoxins from Aspergillus, Fusarium, and other aflatoxin producing organisms [3]. The immunoassay, also a type of immunoaffinity procedure, may be combined with mass spectrophotometry to provide ultrafast immuno-extraction for microanalytical assessments [4]. Many times the assay can be designed for direct detection; however, it is also common for assay design to employ a competitive binding step, in which a signal is generated as the antigen competes with a labeled species for antibody binding sites. The heart of the chromatographic matrix is a conjugated support containing an immobilized antibody with specificity for the antigen. Secondary conjugates are also employed, using molecular pairs such as biotin and streptavidin; Protein A or protein G may also be used, which takes advantage of the Fc component within the antibody molecule. Multiple commercially available supports are currently available for immunoaffinity chromatography use in the laboratory.

The procedure for assessment is relatively straightforward. Samples for detection are applied to immunoaffinity columns in appropriate binding buffers. Affinity binding with the antibody occurs; bound antigens are retained to the column while nonbinding components pass through unimpeded. The antigen is later eluted by disrupting the antibody–antigen complexed interaction with an appropriate elution buffer. Elutions may be accomplished in one step, or by gradient, depending on next steps for antigenic quantitation. Immuno-extraction can often be coupled with a second analytical method.

IMMUNO-ASSESSMENT FOR SPECIFIC PATHOGENS

In many cases, the tests described above may be useful for detection and quantification of pathogenic organisms recovered from clinical samples. Table 11.1 provides a representative sampling of common pathogens where immunologic evaluations of microbial infections are effective. In many cases, confirmation with culture is warranted/or required.

GENERAL TECHNIQUES FOR ANTIBODY RECOGNITION

Immunochemical-based techniques also permit detection of antibodies within the laboratory setting, allowing sensitive and selective

Table 11.1 Representative antigen testing for specific pathogen classes

Pathogen class	Organism	Specimen type	Additional comments
Bacteria	*Streptococcus pneumoniae*, *Neisseria meningitidis*, *Haemophilus influenzae*, GBS	CSF	Low sensitivity and specificity
	GAS	Throat swab	Rapid detection streptococcal pharyngitis
	β-Hemolytic streptococci	Colonies growing on agar plate	Useful for Lancefield typing
	Salmonella	Colonies growing on agar plate	Useful for *Salmonella* subtyping
	Legionella, *Streptococcus pneumoniae*	Urine	Diagnosis disseminated disease
	Treponema	Blood	Nonspecific test for diagnosis syphilis
Viral	Adenovirus	Stool	Diagnosis respiratory or GI disease
	Influenza, RSV	Nasopharyngeal swab	Rapid diagnosis for patients in ER
	Rotavirus	Stool	Culture not available
	HIV	Serum	P24 antigen part of fourth generation test
Fungi	*Cryptococcus*	CSF, serum	Titers useful for tracking treatment
	Histoplasma, *Coccidioides*, *Blastomyces*	Urine, serum, BAL, CSF can also be tested	Diagnosis disseminated disease
	Aspergillus	Serum, BAL	Galactomannan nonspecific, but useful diagnosis disseminated disease

(*Continued*)

Table 11.1 Representative antigen testing for specific pathogen classes (Continued)

Pathogen class	Organism	Specimen type	Additional comments
Parasites	*Cryptosporidium, Giardia*	Stool	EIA is more sensitive than microscopy
	Malaria	Blood	Microcopy common, antigen testing useful when no expertise in smear reading
	Toxoplasma	Serum	TORCH evaluation

TORCHS, Toxoplasma, Rubella, CMV, HSV, and Syphilis.

determination of immune responses induced in the presence of pathogen. Assessment of levels of specific antibodies and antibody isotypes can reveal much about the time course of pathogenic interaction with the patient. In addition assessment of specific antibody response can reveal much about the maturation of immune response.

It is important to understand that secretion of antibodies only occurs after antigenic stimulation of membrane anchored immunoglobulin on the B cell surface. The process requires both antigenic recognition and help given in the form of T cell cytokines. During differentiation, processes occur to change the biological properties of the secreted antibody. The first is class switching, or isotype switching, where antibodies switch from IgM and IgD to another isotype (IgG, IgA, or IgE). Detection of class switching, specifically from IgM to IgG, can be a powerful clinical tool that signifies exposure to pathogens. This differentiates recent infection (IgM) from long-term infectious presence (IgG). Likewise, maturation of response can be identified with increase in relative titers of IgG which is usually indicative of active encounters with the cognate antigen or pathogen (fourfold increase in titer between acute and convalescent samples indicates infection). Finally, functional power of reactivity and specificity of the antibody can be assessed, with both affinity (direct binding constant) and avidity (strength of overall reactivity with antigen) measured. Below we review some of the technology underlying main antibody detection assays used frequently in the clinical laboratory.

One additional important concept to keep in mind in the laboratory setting includes detection of responses in infants. While the fetus begins to produce IgM at 6 months gestation; the majority of IgG at birth is transplacental and maternal in origin. Indeed, infants do not respond well to polysaccharide antigens prior to 2 years of age.

Enzyme-Linked Immunosorbent Assay

A general approach to diagnosing diseases utilizes an immunoassay which indirectly tests for the presence (or change in presence) of antigens by looking for antigen-specific antibodies [5]. These same methods may also be useful to directly test for specific antibodies. Basically the enzyme-linked immunosorbent assays (ELISAs), also known as enzyme immunoassays (EIA) or solid-phase immunoassays, are tests designed to detect antibodies or antigens by producing an enzyme triggered color change. ELISAs are referred to as solid-phase assays because they require the immobilization of antigens or antibodies on solid surfaces. The noncompetitive ELISA utilizes a specific antigen attached to a solid-phase surface (plastic bead or microtiter well). The test specimen is added; incubation allows antibodies in the test specimen to bind to the antigen. A secondary enzyme-labeled antibody is used to detect the test antibody. A chromogenic substrate is added; the enzyme converts the substrate to a color detectable compound. The amount of color that develops is proportional to the amount of antibody in the test specimen. In a variation of detection modalities, fluorescent type of secondary reporter conjugates are used, with measurement of bound fluorescence as the assay endpoint.

A similar assay employs an antigen-specific antibody bound to the well. The test specimen is added, this time analyzing for the presence of antigen. A secondary antibody captures the antigen in a "sandwich" format. Similar to the noncompetitive ELISA, the amount of color (or fluorescence) produced in the presence of enzyme-labeled reporter is proportional to the antigen in the test specimen.

Automated Multiplex Immunoassay

Traditional ELISAs detect and measure antibodies reactive to a single antigen per plate, and are thus limited to sample volume restrictions. Newer methodologies utilize similar principles as described above for the ELISA, by building on the process using uniquely identifiable beads. These beads allow for the simultaneous detection of multiple substances in a single reaction. By adhering capture antibodies to beads, a greater surface area (and flexibility) is garnered, thus allowing multiple assessments in a single well. These newer bead-based multiplex immunoassays can yield a wealth of information on the presence of reactive antibodies and isotypes, or even multiple proteins and biomolecules, in clinical samples, and are useful diagnostic assays for both identification and assessment of progression of disease states. In addition, automation now allows considerable reduction in overall assessment time.

One example multiplex immunoassay is the Bio-Plex Assay, commercially available from Bio-Rad. In their assay, magnetic beads are colored internally with two different fluorescent dyes (red and infrared), to generate up to 100 distinct bead regions. Each bead region is conjugated to a specific target analyte or antigen. Subsequent binding with a biotinylated detection antibody occurs. Utilization of a reporter dye, such as a streptavidin-conjugated phycoerythrin, permits quantitative assessment.

Immuno-Diffusion

The no longer used Ouchterlony assay, developed in the 1950s, represents the historical model for characterization of precipitin reactions that fall into the category of immune-diffusion (ID) assessment. Simply put, in an ID assay, applied antigen diffuses through a gel matrix containing reactive antibodies; as antigen and antibody meet, the diameter of the precipitin ring formed can be quantitated by comparison to known standard antigen concentrations. A variation of this, the radial ID assay, allows quantitative measurement of antigen. Specifically, antibody is directly incorporated into the assay gel matrix. This permits qualitative determination as antibody and antigen diffuse toward each other through the matrix. A precipitin ring forms as the antigen and antibody interact; the amount of antigen applied to the well is proportional to the diameter of the precipitin ring. Comparison of unknown samples to known antigen concentrations allows quantitative analysis.

Western Blots (Immunoblotting)

The Western blot technique represents a qualitative measurement of antibodies reactive to specific antigens. In this case, a mixture of proteins is separated by electrophoresis through a gel matrix, and then transferred onto a solid medium such as nitrocellulose that binds proteins tightly. Antibodies are incubated with proteins bound to the nitrocellulose.

Coupled molecules detect bound antibody, using reporter enzyme covalently attached to detectors. Substrate for the enzyme is added, which turns colors when enzyme is present. The resulting colored line demonstrates that the antigen was present in the initial population of mixed proteins. An alternative visualization can be done using fluorescent tags. This type of test is clinically useful for detection of patient antibodies reactive to viral antigens, such as the *gag*, *pol*, or *env* gene products produced by the HIV virus. Care should be taken for false negatives, especially if the patient is immunocompromised.

Complement Fixation

The complement fixation test is a blood test in which a sample of serum is exposed to a particular antigen and complement in order to determine whether or not antibodies to that particular antigen are present. The nature of complement is to react in combination with antigen–antibody complexes. The relative lack of antigen specificity allows complement to react with almost any antigen–antibody complex. In the test procedure, complement remains free unless fixed by the particular antigen and antibody system in question. The indicator used in many complement fixation assays is sheep RBCs. In a positive or reactive test, the complement is bound to an antigen–antibody complex and is not free to interact with target RBCs. The RBCs remain unlysed and settle to the bottom of the well to form a button. In a negative or nonreactive test, the complement remains free to interact with the blood cells causing them to lyse. The complement test is a powerful tool to identify antibodies reacting with antigens, and it permits confirmation of exposure to a specific microorganism. For example, the Wasserman reaction is an example of a diagnostic complement fixation test to detect antibodies to the syphilis organism Treponema; a positive reaction indicates the presence of antibodies and therefore syphilis infection.

SPECIFIC PATHOGEN IMMUNE ASSESSMENT

In many cases, the tests described above may be useful for detection and quantification of antibodies directed toward pathogenic organisms. Table 11.2 provides a representative sampling of common occurrences where immunologic evaluation of antibodies directed toward microbial epitopes is effective in clinical diagnostics. It is critical to understand that there are limits to use of antibody detection in our understanding of

Table 11.2 Representative antibody testing in common laboratory diagnostics

Pathogen class	Organism	Specimen type	Additional comments
Bacteria	*Borrelia*	Serum	Diagnosis lyme disease EIA and confirmation WB
	Treponema	Serum	Confirmation of Syphilis (EIA, VDRL, TPPA, FTA ABS)
	Brucella, Bartonella	Serum	Diagnosis atypical, slow growing, rare organisms
Viral	HIV	Serum	EIA screen, IFA or WB confirmation
	CMV, HSV	Serum	Useful in pregnancy (TORCH) and pre transplant test of immunity
	Rubella	Serum	Useful in pregnancy (TORCH) evaluation of immunity
	West Nile virus, Zika, Arboviruses	Serum	Useful in diagnosis
	EBV	Serum	Useful diagnosis mononucleosis
Fungi	*Histoplasma, Blastomyces, Coccidiodes*	Serum	Can be used in conjunction with urine antigen testing diagnosis disease
	Aspergillus	Serum	Can be used in allergic sinusitis or disseminated disease
Parasites	*Toxoplasma*	Serum	Useful pregnancy (TORCH) evaluation immunity pregnancy, diagnosis encephalitis in immunocompromised patients
	Strongyloides, Filaria, other uncommon blood and tissue parasites	Serum	Presence of antibody indicative of disease organisms uncommon

active infection. Often, false positive IgM can be found due to nonspecific factors like connective tissue diseases. Indeed, it takes time for antibody titers to rise postexposure; for example, a fourfold rise in titer between acute and convalescent is really necessary for diagnosis of current disease. Therefore, the power in detection may be of more use for epidemiological analysis and diagnosis of uncommon diseases where the majority of the community population has not been exposed to the organism (and therefore have no detectable antibody). Finally, these tests can be useful to prove immune status in pregnant women, although there are limits for utility in infants since most IgG represents maternally derived antibody.

KEY POINTS

- Immune-based assays are designed to assess the presence pathogenic organisms in clinical samples, and are instrumental in our monitoring of clinical progression to disease and subsequent response to therapeutic treatments.
- Immunochemical-based techniques are built upon the ability of antibodies to specifically recognize target shapes (antigens) unique to pathogenic molecules (antigens). They offer the advantage of specific antigen recognition, and permit sensitive and selective determination of target factors present in clinical samples.
- Precipitation reactions are used for analysis in situations where only qualitative detection is required. Quantitative assessment usually requires higher sensitivity in detection of complexes, with subsequent comparisons to known standards.
- Agglutination reactions may be accomplished using latex (synthetic) particles, and typically permit rapid assessment with consistency, uniformity, and stability.
- The ELISA has given rise to second generation multiplex immunoassays, which are useful as diagnostic tools for multiple antigens present in small clinical samples. Automation in the laboratory setting has reduced detection time, while maintaining and/or increasing sensitivity.
- Rapid chromatographic immunoassays are immunoaffinity procedures that may be combined with mass spectrophotometry to provide ultrafast immuno-extraction for microanalytical assessments. They have become extremely useful for rapid isolation of specific antigenic factors from specimens.

REFERENCES

[1] Kohler G, Milstein C. Continuous cultures of fused cells secreting antibody of predefined specificity. Nature 1975;256(5517):495–7.
[2] Koczula KM, Gallotta A. Lateral flow assays. Essays Biochem 2016;60(1):111–20.
[3] Anfossi L, Giovannoli C, Baggiani C. Mycotoxin detection. Curr Opin Biotechnol 2016;37:120–6.
[4] Moser AC, Hage DS. Immunoaffinity chromatography: an introduction to applications and recent developments. Bioanalysis 2010;2(4):769–90.
[5] Actor JK. Elsevier's integrated review immunology and microbiology. Philadelphia, PA: Elsevier; 2011.

CHAPTER 12

Overview of Molecular Diagnostics Principles

Contents

INTRODUCTION

With the success of understanding molecular basis of many diseases and pathogens, nucleic acid-based testing is becoming a valuable diagnostic tool not only in the setting of inherited genetic disorders such as cystic fibrosis but also in diagnosis of many hematological, neoplastic, and infection processes. Molecular diagnosis is also applicable in the area of personalized medicine where understanding genetic mutations of genes encoding various liver enzymes that metabolize drugs help clinicians to decide what particular drug at what dosage should be appropriate for the patient depending on whether a patient is an extensive metabolizer, intermediate metabolizer, poor metabolizer, or ultrarapid metabolizer.

Molecular detection techniques continue to increase in clinical microbiology laboratories. The implementation of in vitro nucleic acid amplification techniques lead by real time polymerase chain reaction (PCR) has transformed speed and efficacy of viral detection and selected bacterial detection. Although molecular biological techniques may not replace

Microbiology and Molecular Diagnosis in Pathology.
DOI: http://dx.doi.org/10.1016/B978-0-12-805351-5.00012-0

conventional techniques used in clinical microbiology laboratories in the future, but molecular testings play a major role in clinical microbiology laboratories due to commercial availability of many FDA approved molecular tests and some automation [1].

BASIC PRINCIPLES OF MOLECULAR BIOLOGY

A nucleotide, the smallest unit of genetic information, is composed of a deoxyribose (five carbon sugar) with an attached nitrogenous base and a phosphate group. The nitrogenous base can either be a purine (adenine [A] or guanine [G]) or a pyrimidine (cytosine [C] or thymine [T]). Nucleotides attach to each other linearly and form long strands (polynucleotides). These strands serve as the basic genetic code in living organisms and are referred to as deoxyribonucleic acid (DNA).

The human genome in a single cell is estimated to have approximately 6.4 billion nucleotides. With so much genetic information stored in a single cell, it needs to be packaged and stored in an efficient manner. In humans, DNA is packaged in two steps. First, two polynucleotide strands are attached to each other by hydrogen bonds forming a structure we refer to as a DNA double helix. The two strands are complementary to each other and one strand is designated as the forward strand and the other as the reverse strand. A strand is complementary to another strand when the sequence of one strand binds to a specific sequence of another strand. Specifically, an adenine on one strand will only bind to a thymine on the other strand by two hydrogen bonds and the cytosine on one strand will only bind to a guanine on the other strand by three hydrogen bonds. During the second step of DNA packaging, the double helix DNA are wrapped around histone proteins, and these proteins are packed tightly together until they form a chromosome. Each normal human cell contains a total of 46 chromosomes that is made up of two sets of 22 chromosomes and a pair of sex chromosomes. Since a normal human cell has two sets of chromosomes, this is referred as diploid. An abnormal human cell can have different multiples of chromosomes referred to as aneuploidy (e.g., three sets is a triploid and four sets is a tetraploid).

In order for DNA to be inherited, replication has to occur. Replication occurs when DNA strands are copied resulting in identical copies of the original (template) DNA. One important concept to understand is that DNA has directionality where one end of the DNA strands is the 5′ (five-prime) end and one end is the 3′ (three-prime) end. By convention, DNA read from 5′ to 3′, making the strand oriented from 5′ to

3′ the forward strand and its complementary strand the reverse strand. This is important because DNA polymerase adds on additional nucleotides during replication onto the 3′ end and not the 5′ end. By convention, the strand that is being replicated in a continuous fashion is the leading strand and the opposite strand is the lagging strand.

Replication occurs in three steps: Initiation, elongation, and termination. First, during initiation, particular points in the DNA called "origin" are targeted by initiator proteins that unwind and unzip the double stranded DNA molecule. Once a small portion of the DNA is unwound, an ribonucleic acid (RNA) primase attaches to the forward strand in the opposite direction (the 5′ end of the RNA primase will be attached to the 3′ end of the forward strand). Then, the DNA polymerase will add nucleotides onto the 3′ end of the RNA primase making a copy (leading strand) of the forward strand. On the reverse strand, the RNA primase will attach to the reverse strand; however, because of how the replication fork forms, it will not be able to replicate in a continuous fashion. Instead, it will sequentially add on RNA primases on varying points of the reverse strand as the replication fork unfolds. The DNA polymerase will add nucleotides onto the 3′ end of the RNA primase until it reaches the 5′ end of the previously attached RNA primase. Each of these fragments of RNA primase with attached nucleotides (Okazaki fragments) form the lagging strand. There are gaps between each of the Okazaki fragments which are filled in by ligase [2]. In the last step of replication termination, the copying of DNA ends once a termination sequence is read on the template DNA and a protein binds to the sequence to physically stop the replication process.

The central dogma in molecular genetics is that DNA is transcribed into RNA which is translated into proteins. In transcription, a focus or region of a DNA molecule is copied to form a complementary RNA sequence. The RNA can be a coding messenger RNA (mRNA), which will serve as a template for translation of proteins or it can be a noncoding RNA such as a microRNA, ribosomal RNA, transfer RNA, or ribozymes. In transcription, an RNA polymerase binds to a promoter region upstream of the region of interest on the DNA molecule and the RNA polymerase separates the two complementary DNA strands and creates a transcription bubble. The RNA polymerase then adds one RNA nucleotide at a time to one strand of the DNA molecule. An RNA nucleotide is different from a DNA nucleotide in that the RNA nucleotides have a ribose backbone instead of a deoxyribose backbone. Also, while they have the same nitrogenous bases of guanine, adenine, and cytosine, instead of thymine, RNA nucleotides have uracil. Once

RNA strand of the gene of interest are synthesized, the hydrogen bonds of the twisted RNA–DNA helix break apart, which will free the RNA strand from the DNA. The RNA strand is then further processed with an addition of a 5′ cap, addition of a poly-A tail, and splicing. A 5′ cap is a single G that is added to make the 5′ end appear as a 3′ end, which serves to protect it against exonuclease (enzyme that targets and degrades the 5′ end). Also, the 5′ cap can be recognized by a nuclear core complex that allows the mRNA to leave the nucleus. Polyadenylation occurs when multiple adenines are added to the 3′ end of an RNA transcript and this is important for stability of the mRNA, nuclear export, and translation. Within the RNA strand are coding (exon) and noncoding regions (introns). During splicing, the introns are removed from the RNA strand and the remnant exons are joined back together forming the final mRNA product to be translated into a protein.

During translation, the genetic code on the mRNA that was produced during transcription is read and translated an amino acid chain (polypeptide). This polypeptide folds into a protein that is able to perform the various functions in a cell. Translation occurs in four steps: initiation, elongation, translocation, and termination. In the first step, initiation, a ribosome attaches to the mRNA and an initiator transfer RNA (tRNA), connected to the amino acid methionine attaches to the start codon site. Next, in elongation, a tRNA adds an amino acid that corresponds to the mRNA codon next to the start codon onto the previous amino acid. Then, the ribosome moves onto the next mRNA codon and elongation continues, until termination occurs when a stop codon is reached and the ribosome releases the polypeptide. Amino acids are the basic unit of protein and contain an amine group, carboxylic acid group, and a side chain specific to each amino acid. There are 21 total amino acids including arginine, histidine, lysine, aspartic acid, glutamic acid, serine, threonine, asparagine, glutamine, cysteine, selenocysteine, glycine, proline, alanine, valine, isoleucine, leucine, methionine, phenylalanine, tyrosine, and tryptophan. Each amino acid is coded by one or more triplicate of specific nucleotides. For example, tyrosine is coded by the UAU or UAC and serine is coded by UCU, UCC, UCA, or UCG. Some other notable amino acids are the start codon methionine which is coded by the nucleotides AUG and the stop codon which is coded by UAA, UAG, or UGA.

NOMENCLATURE IN MOLECULAR DIAGNOSTICS

Nomenclature can be a subject of confusion in molecular diagnostics. This section describes three nomenclatures commonly used in molecular diagnostics: Coordinates of a nucleotide, cytogenetic nomenclature, and mutation nomenclature.

Coordinates of a Nucleotide

Before delving into the subject of naming the position of a nucleotide, some background of the human reference genome will be discussed. The human genome project was started by the United States in 1990 and with the help numerous international universities and research centers it was completed in 2003 [3]. It was completed with map-based sequencing using Bacterial Artificial Chromosome (BAC). First, the human genome is broken down into fragments of 150,000–200,000 base pairs in length and cloned in bacteria. These BAC clones are then broken down into smaller fragments that are about 2000 base pairs and sequencing reactions are then carried out on these subclones. The sequences of the subclones can then be mapped to the BAC and the BAC can then be mapped to the location on the human genome. Currently, the genome reference consortium updates the reference genome for humans and other organisms regularly. The latest genome assembly released is the GRCh38 in 2013 [4]. Understanding that different versions of the human genome are available is very important when referring to a position of a nucleotide because the position can be different based on which version of the human genome is used.

Once the version of the reference human genome is determined, a specific position on a sequence for a particular reference sequence can be determined. To write out the position number, start with which chromosome it is on followed by a colon (:) mark and the position number on the chromosome. For example, the *APC* gene based on the GRCh38 assembly is located on chr5:112707505-112846239. With this gene's coordinates, a genome browser like UCSC (https://genome.ucsc.edu/) can be used to examine the reference gene and to get a multitude of information (mapping, sequencing, gene prediction, phenotype, literature, regulation, etc.) regarding this genetic location.

Cytogenetic Nomenclature

The International System for Human Cytogenetic Nomenclature (ISCN) is the central reference used for the nomenclature of human chromosomes. First established in the 1960, it has had many updated versions. Its latest version, ISCN 2013, was published in November 2012 [5]. In cytogenetic nomenclature, the overall number of chromosomes, followed by the sex chromosomes, affected chromosomes with type of abnormalities described in shorthand, location on chromosome, and lastly, the number of cells (in brackets) with a given karyotype is written. The most common structural abnormalities described in shorthand include deletion (del), inversion (inv), and translocation (t). Some of the less common types of abnormalities include: Derivative chromosome (der), dicentric chromosome (dic), duplicate (dup), insertion (ins), isochromosomes (i), marker chromosome (mar), and ring chromosome (r). The long arm (q) or the short arm (p) is first designated to name the location on a chromosome. One easy way to remember that the short arm is "p" is that "p" stands for petite. Next, the p or q will be followed by the number of the region, band, and then a period followed by the subband.

An example is "46,XX,t(9;22)(q34.1;q11.2)" in a Philadelphia chromosome positive patient. In this example, she a patient with 46 chromosomes and is female based on the designation by the two X chromosomes. If it were a male patient, it would be designated by XY instead. In this patient, a translocation between chromosomes 9 and 22 is present, represented by t(9;22). The break is occurring in region 3, band 4, and subband 1 on the long arm of chromosomes 9 and region 1, band 1, and subband 2 on the long arm of chromosome 22. An example of a deletion would be "46,XX,del(5p)," which represents a female with *cri du chat* syndrome that has a deletion in the short arm (p) of chromosome 5.

Mutation Nomenclature

Just as ISCN publishes guidelines on how to describe cytogenetic findings, the Human Genome Variation Society (HGVS) publishes guidelines (http://www.hgvs.org/) on mutation nomenclature. Before discussing the HGVS guidelines regarding nomenclature, the standards for gene symbols and how to find a reference sequence to base mutation nomenclature by will be reviewed.

The Human Genome Organization (HUGO) Gene Nomenclature Committee (HGNC) website (http://www.gene.ucl.ac.uk/nomenclature/index.html) should be referenced to determine the standard gene symbol used to name the genes. On this website, the database can be searched by gene names to find the correct symbols for the gene. For human gene symbols, it should be all *CAPITALIZED* and *ITALICIZED*. An example is "*CFTR*," the symbol for cystic fibrosis transmembrane conductance regulator gene that is mutated in cystic fibrosis patients. Human proteins should be all CAPITALIZED but not ITALICIZED. Lastly, for genes of mice and rats it should be in *lower case* and *italicized*.

In order to accurately describe a mutation when compared to a reference sequence, a specific reference sequence should be identified. In the previous discussion regarding the coordinates of a nucleotide, it was mentioned that reference sequences are constantly updated. This is why it is important to specify which reference sequence is being referred to when naming a mutation because the position of the mutation could be different based on which reference sequence is referred to. To search for the reference sequence, an online tool developed by National Center for Biotechnology Information (NCBI) called NCBI Basic Local Alignment Search Tool (BLAST) on the website http://blast.ncbi.nlm.nih.gov/Blast.cgi can be used. The BLAST tool can help align proteins, mRNA, or genes with reference material to compare sequences, make primers, and find genes. A search of a specific sequence on the web tool can be performed and it will give names of reference sequences that can be referred to when naming a mutation. "NM_000492.3" is an example of a name for a particular GenBank reference sequence for *CFTR*. Reference sequence names that start with NM_ are reference sequences for mRNA, ones that start with NP_ are reference sequences for proteins, and ones that start with NG_ are reference sequences for genomic DNA.

Now, after reviewing gene symbols and reference sequences, how to name a mutation with HGVS guidelines will be reviewed. A mutation will be named by starting off with the GenBank reference sequence used, followed by a colon (:), the nucleotide number, and the mutation in question. An example for a *CFTR* missense mutation would be "NM_000492.3:c.3909C > G," where using the GenBank reference sequence of NM_000492.3, a substitution (change) from C to G at position 3909 is present. This change would cause a protein change of asparagine to lysine at amino acid 1303 (p.Asn1303Lys). The symbol > means change, c. represents a coding DNA sequence, and p. represents a protein sequence.

BASIC TECHNIQUES USED IN MOLECULAR DIAGNOSTICS

One of the major purposes in molecular diagnostics is identifying aberrations in the genetic code and a multitude of methodologies can be used for this. A molecular technique is chosen based on the genetic aberration that is to be detected. In general, a cytogenetic technique such as conventional karyotyping, virtual karyotyping, or fluorescent in situ hybridization (FISH), gives an overview of the entire DNA genome, analogous to having an aerial overview of a forest. PCR techniques give information on specific parts of the DNA genome, analogous to looking at individual trees in a forest. Lastly, sequencing techniques gives a detailed view of the DNA genome, analogous to data on how many leaves or branches are on an individual tree. Before going into detail about some of the more common molecular techniques, sample collection for molecular testing will first be discussed.

Specimen Considerations

All the cells in the human body have DNA material, and so dependent on what the molecular test is trying to identify, a specific type of tissue will be obtained. For example, to identify stool pathogens in the stool, a stool sample can be obtained. If mutations in a gastrointestinal tumor are in question, a piece of tissue from the bowel with the tumor will be obtained. Besides considerations of the type of tissue that is obtained, other considerations include the concentration of the tissue and how the tissue is stored.

It is important to consider the concentration of the tissue involved because a minimal amount of DNA is required for an assay. If the minimal amount of DNA is not obtained, there is a higher chance of having a false negative result. So, for many assays, one important step is to measure the amount of DNA extracted from the sample before it is run on the assay. Tumor heterogeneity is another important concept to keep in mind when interpreting molecular results. If the sample has a low tumor percentage, the confidence that a negative result is truly negative is low because the results can be due to the normal tissue instead of the abnormal tissue. When working with paraffin embedded tissue of tumor samples, it is of utmost importance to examine the histologic slide for tumor percentage. The minimum tumor percentage required for an assay is determined by each institution based on their experience and validation of an assay. Techniques such as macrodissection or lesser microdissection of paraffin embedded tissue can be used to increase the tumor percentage of the sample that is sent for molecular testing.

Tissue can be fresh, frozen, or preserved in a chemical solution. How the tissue is stored can affect the DNA quality and hence the results of the molecular testing. In surgical pathology, tissue is usually stored in a fixative which has many advantages such as stabilizing and strengthening the tissue and disabling proteolytic enzymes so that the tissue can be processed and stained for histological analysis under the microscope. However, certain fixatives can degrade the DNA or RNA more readily than others. Two main types of fixatives are used, cross linking fixatives (e.g., formaldehyde and glutaraldehyde) and precipitating fixatives (e.g., alcohol and methanol). In general, the DNA and RNA quality of tissue that has been fixed in cross linking fixatives is poor while the tissue that has been fixed in precipitating fixatives is better. However, in terms of fixatives in formalin, the DNA and RNA quality can be fair if fixed in buffered formalin as opposed to unbuffered formalin.

Cytogenetic Karyotyping

Cytogenetic karyotyping is a cytogenetic technique that allows for an overview of cytogenetic abnormalities such as aneuploidy (changes in chromosome numbers), translocations, inversions, and deletions of the patient's chromosomes. In classic karyotyping, cells from various sources such as blood, bone marrow, tumor, or amniotic fluid is cultured to grow on cell plates and then a mitotic inhibitor such as colchicine is added to the cell culture to stop the mitosis at metaphase. After the inhibitor is removed, the cells are placed in a hypotonic solution to lyse the red blood cells; and more importantly, the white blood cells will swell and allow for the individual chromosomes to spread out for analysis. The cells are next fixed and stained with different types of stains and then arranged by a cytogenetic specialist. The human chromosomes are arranged in pairs from chromosome 1 (largest) to chromosome 22 (smallest) with the sex chromosomes at the end. The short arm of the chromosome is oriented toward the top and the long arm is oriented toward the bottom. After being analyzed by a cytogenetic specialist, it is reviewed by a board-certified cytogeneticist who issues the final cytogenetic report after taking into account the clinical history of the patient.

Many different stains can be used to stain the chromosomes resulting in different banding patterns of the chromosomes. The stain used depends on what the karyotyping is trying to identify. Currently, the most widely used banding technique is G-banding. Here, the chromosomes are digested by trypsin before staining it with a Giemsa stain creating lightly (GC rich regions) and darkly (AT rich regions) stained bands on the chromosomes. Each individual chromosome has a characteristic banding pattern, so cytogenetic changes can be identified based on changes in the banding pattern. Some other banding techniques include the Q-banding (stains with quinacrine that cause a fluorescent pattern similar to G-banding), R-banding (reverse of G-banding where AT rich regions stains lightly and GC rich regions stain darkly), C-banding (stains centromeres), and T-banding (preferentially stains telomere regions).

Fluorescent In Situ Hybridization

FISH is a cytogenetic technique that uses fluorescent-labeled probes to detect the absence or presence of specific regions/sequences on chromosomes. In FISH, a cytogenetic slide preparation or another type of slide preparation (bone marrow smear, paraffin embedded tissue, cytology-spin preparations, blood smear, etc.) is aged in a salt solution, dehydrated with ethanol, and then fluorescent probes are added. The fluorescent-labeled probes are designed to be complementary to sequences, and hence, regions of interest on the chromosome. The DNA on the slide preparation and the probes are heated allowing both of them to denature and then the preparation is cooled so they are allowed to anneal to each other. After the unbound probes are washed off, it is counterstained with 4′6-diamidino-2-phenylindole (DAPI). The slides are analyzed by a cytogenetic specialist where 200–1000 cells are examined before finally being reviewed by a molecular pathologist or a board-certified cytogeneticist.

Many different types of probes are available to detect different alterations on chromosomes. In a basic FISH probe, a single probe can be used to detect presence or absence of a specific sequence (a mutation, a deletion, etc.). Another type of FISH probe is called a break-apart probe where if there is no translocation involving the area the probe is attached to, there will be one signal and this will be considered a negative result. However, when there is a translocation occurring in the area the probe anneals to, the translocation causes a break in the middle of the probe causing two separate signals to result and this will be considered positive result. This is useful for examining genes like *EWSR1* that has many fusion partners. When the *EWSR1* probe breaks apart into two signals, it means that a fusion has occurred; however, by this test alone the fusion partner is not known. Another type of commonly used probe is a fusion probe, where two probes with different fluorescent colors sit on two different chromosomes. When a fusion occurs, the two probes will come together and result in just one signal instead of two which signifies a fusion of the two chromosomes. A common fusion probe is for the *BCR-ABL1* transcript where a probe sits on the *ABL1* on chromosome 9 and another probe of another fluorescent color sits on *BCR* on chromosome 22. In cells with a *BCR-ABL1* transcript, these two probes will come together allowing us to identify the fusion transcript in the cells.

Virtual Karyotyping

A relatively newer technology in the realm of cytogenetic is virtual karyotyping. In virtual karyotyping, copy number changes such as gains, deletions, and amplification can be identified at a higher resolution (10–20 kilo-bases) when compared to conventional karyotyping (5–10 megabases). While a higher resolution can be achieved by virtual karyotyping, it cannot identify balanced translocations, inversions, and ringed chromosomes because there are no copy number changes. Therefore, in its current form, virtual karyotyping serves as an adjunct to traditional cytogenetic techniques and not as a replacement.

Conceptually, virtual karyotyping is performed by hundreds to millions of probes that are attached to an array hybridizing to fragmented DNA (the fragments are from both normal DNA and DNA from tissue of interest). The ratio of the hybridization signal between DNA of interest and normal DNA is captured by the virtual karyotyping hardware and since the location of the probes on the chromosomes is known; the software is able to reconstruct in silico the chromosomes visually on a computer. Being able to visualize the ratio between the normal and the tissue in question, the pathologist interpreting the virtual karyotype is able to determine if copy number changes are present.

Two main types of virtual karyotype, array-comparative genome hybridization (aCGH) and single nucleotide polymorphism (SNP) array, are used. SNP-arrays have come into vogue due to the added advantage of being able to detect copy number neutral loss of heterozygosity. The current leader in the SNP array space is Affymetrix, where they have an OncoScan assay for formalin fixed-paraffin-embedded tissue (FFPE) (900k to 1 million probes) and a CytoScan assay for blood with 3.2 million probes.

Polymerase Chain Reaction

PCR is a technique that was developed in the 1970s and has revolutionized molecular diagnostics. The basic goal of PCR is to amplify many copies of a target DNA in order to examine that particular focus of DNA more closely. This is done through multiple thermal cycles that consist of three steps: Denaturing, annealing, and extension; and followed by a final elongation and hold step. In a basic PCR, the template DNA with the target DNA, primers, DNA Polymerase, deoxynucleoside triphosphates (dNTP), buffer solution, monovalent cation such as K^+, and bivalent cation such as Mg^{2+} or Mn^{2+} are placed into a PCR tube. These reaction tubes are then placed into a thermal cycler that will cycle at different temperatures and times based on what the operator programs the thermal cycler to do.

In the first step, denaturing, the thermal cycler is heated to 94–98°C for 20–30 seconds, which causes a disruption of the hydrogen bonds between the two DNA strands and results in two individual DNA strands. Next, in the annealing step, the temperature is decreased to 50–65°C for 20–40 seconds to allow annealing of the primers to the single DNA strands. The primers are designed to anneal to regions flanking the 5′ and 3′ end of the target sequence. During last step of the thermal cycle, the extension step, the thermal cycler is heated to 72°C in order for the DNA polymerase to add the phosphate group on the 5′ end of dNTPs onto the hydroxyl group on the 3′ end of the primer (after the first cycle it will be adding dNTPs to the hydroxyl group of the newly synthesized strand). These cycles are performed 20–40 times dependent on how many copies of target DNA is desired and is limited by the amount of reagents present in the reaction. Each of the cycles doubles the amount of target sequence if the reaction efficiency is 100%. So theoretically, at 30 cycles and 100% efficiency, one original template DNA sequence will have greater than 1 billion copies ($2^{30} = 1.07 \times 10^9$). Once the desired number of cycles is ran, a final elongation step occurs by adjusting the temperature to 70–74°C for 5–15 minutes after the last PCR cycle. Lastly, the temperature is reduced to 4–15°C for short-term storage. The product of the PCR, which is basically multiple copies of a target (desired) DNA sequence, has many utilities. Furthermore, many variations of the basic PCR have been developed.

Design of Polymerase Chain Reaction

When designing a PCR, three main things needs to be considered: Primer design, reagent concentrations, and thermal cycling conditions.

Primers are sequences that are usually 18–30 nucleotides in length that bind to the template sequences being replicated during PCRs. The primer sequences usually flank the 5′ and 3′ end of the target sequence and attach to the template DNA during the annealing step of the PCR. The primer sequences will be complementary sequences to the flanking regions of the target DNA and the complementary areas are selected based on a few guidelines in order for the primers to be able to anneal to the template sequence in a specific and efficient manner. Some considerations for primer selection include self-complementary, GC content, and melting temperature (Tm) of the selected primers. The primers for forward and reverse sequence cannot be self-complementary to each other because if they are self-complementary the two primers will anneal to each other and result in primer dimers instead of annealing to the target DNA. The GC content (percentage) is the number of GC nucleotides divided by the total nucleotides. GC content should be similar for both the forward and reverse strands, ideally between 40% and 60%, in order for the reaction to run efficiently. Lastly, the Tm for both the forward and reverse primers should be between 65°C and 75°C and within 2–4 degrees of each other so that the PCR can run efficiently. The Tm is calculated by the formula $[(A + T)2 + (G + C)4]$, where A is the number of Adenine, T is the number of Thymine, G is the number of Guanine, and C is the number of Cytosine in the primer sequence. So, longer primers have stronger bond between the primer and target sequence and therefore a higher Tm. Also, primers with higher GC content have more hydrogen bonds as compared to primers with higher AT content (three hydrogen bonds vs two hydrogen bonds), resulting in primers with high GC content having higher Tm's. Another way to view Tm is that primers with lower Tm bind to other DNA sequences in a less specific manner when compared to primers with a higher Tm. When PCR was first developed, researchers would design primers by paper and pencil taking into account the aforementioned guidelines to design the most optimal primers for a PCR. However, now many online tools are available where computer algorithms can help design the most optimal primers for a desired PCR. One popular tool is on the NCBI website called Primer-BLAST (http://www.ncbi.nlm.nih.gov/tools/primer-blast/).

The second consideration in PCR design is the amount of regents used in the reaction. Referencing the literature on an assay that has already been well validated or performing experiments with different concentrations of reagents can be done to find the most optimal amount of reagents to use. One reagent component that is especially important in the design of a PCR is the bivalent cation (salt) such as the Mg^{2+}. Mg^{2+} acts as a stabilizer of double stranded DNA and too much Mg^{2+} will allow nonspecific binding of primers to DNA and also prevent complete denaturing of double stranded DNA. One the other hand, Mg^{2+} also acts as a catalyst for Taq polymerase, and so too little Mg^{2+} will cause the Taq polymerase to not function properly. Hence, it is important to find the most optimal concentration of Mg^{2+} while balancing these two factors when trying to find the optimal PCR conditions.

The last category to consider is the thermal cycling conditions, which like the quantity of reagents used can be determined by referring to previously published literature or can be determined by experimental studies. The operator of the thermal cycler will determine the exact cycling conditions of each step of the cycle and determine how many cycles to run. One especially important cycling condition is the annealing step. Here, the temperature must be low enough for hybridization of the strand to occur but also high enough for the hybridization to be specific. If the annealing step is set at the wrong temperature, an error will result either due to nonbinding of the primer to the template strand or nonspecific binding between them.

Applications of Polymerase Chain Reaction

PCR has a multitude of applications and its applications can be categorized in two different ways. The first way is to base it upon the purpose of the PCR. The three main purposes are isolating the target DNA, quantification of the DNA, and disease diagnosis. In DNA isolation, the amplified DNA can be used as hybridization probes for Southern or Northern blot, it can be sequenced with different sequencing technologies to identify the sequence of the DNA, and the isolate can also be run on electrophoresis to be used in purposes such as paternity testing. Quantification can be done with techniques such as quantitative PCR (qPCR) to determine the quantities of the sequence in question. Lastly, disease diagnosis can be performed in oncology by detection of specific translocations. This can be performed by detection of the translocation product from the PCR by electrophoresis or by amplifying segments of sequences that are known to harbor mutations and sequencing the PCR products to identify such mutations. It can also be used in the diagnosis of infectious disease through identification of organisms by sequencing or electrophoresis techniques.

Another way to categorize the application of PCR is to categorize it based on its medical applications, infectious disease applications, forensic applications, and research applications. Medical applications include genetic testing, tissue typing, and detection of alterations to oncogenes. Infectious disease applications include identification and quantification of bacterial, viruses, and/or parasites. Forensic applications include genetic fingerprinting and paternity testing. Research applications include DNA cloning, DNA sequencing, gene expression, gene mapping, and creation of hybridization probes for other molecular techniques such as Southern and

Northern blot. The application of PCR as well as other techniques will be discussed in detail in the next chapter.

Variations of Polymerase Chain Reaction

So far, the basic concepts behind PCR have been discussed; however, over the past few decades, many different variations of the basic PCR technique have been developed. Here, a few of the more commonly used variations of PCR will be discussed, including qPCR, multiplexed PCR, RT-PCR, HOT-start PCR, and nested PCR.

Quantitative Polymerase Chain Reaction

qPCR, also known as real time PCR, is a variation of the PCR where instead of simply having multiple amplified target sequences to analyze at the end of the PCR run, this technique has the ability to quantify the amount of gene product in the sample during the PCR cycling reaction. qPCR can be used in a qualitative (determine whether a DNA product is present), semiquantitative (determine whether the amount of target DNA is more or less than a certain threshold), or quantitative (actually measuring the amount of target DNA) manner. qPCR can also be reported in terms of absolute versus relative quantification. In absolute quantification, the curve generated by the qPCR is compared to the qPCR curve generated by DNA standards, and the curve will give an absolute copy number of the targeted sequence during each cycle of the qPCR. Relative quantification is performed by comparing the quantity of a target sequence to a control reference sequence (otherwise known as a housekeeping gene) that is run at the same time as the target sequence on the qPCR. An example of this is illustrated by an assay that detects *BCR-ABL1* fusion transcript with *ABL* as the housekeeping gene. The results for this assay can be reported as "x" copies of *BCR-ABL1* per "x" copies of *ABL*. If 100 copies of *BCR-ABL1* transcripts were detected per 1000 copies of *ABL*, than it would be 100/1000 or 10%. One important value to know is the cycle threshold (Ct) value, which is the number of cycles of PCR it takes for the fluorescent signal to cross the threshold in order to call the reaction positive. A low Ct value means that there is more target nucleotide in the sample and a high Ct value means that there is less.

In qPCR, after every PCR cycle, sensors in the machine will detect fluorescence from production of additional DNA products and plot the fluorescence on a curve giving the relative quantity of transcript product which can be compared to either DNA standards or a housekeeping gene as described above. A double stranded DNA binding dyes method or a fluorescent reporter probe method is usually used. In the doubled stranded DNA binding dye method, a DNA dye binds to all double stranded DNA and so the more double stranded DNA product is present, the stronger the fluorescence during the qPCR. One advantage of this methodology is that only one primer pair is needed for the reaction; however, nonspecific binding of the dye to double stranded DNA sequences that are not the desired target sequence can occur. With the double stranded DNA binding dye method, the qPCR machine is also able to identify the specific DNA fragment with melting curve analysis. Here, the qPCR machine obtains the Tm of the target sequence, which is specific to the target sequence.

In the fluorescent reporter probe method, a sequence of DNA complementary to a portion of the desired target sequence has a reporter (that fluoresces) on one end a quencher (that absorbs the floresence of the reporter when it is close to the reporter) on the other end. During a PCR, the probe is added to the PCR mixture and during the annealing phase, the probe and primers both attach to different sections of the target sequence. As the DNA polymerase starts attaching nucleotides and forms a new strand, its exonuclease breaks down the probe and separates the quencher from the reporter. This allows the reporter to emit its fluorescence and to be detected by the qPCR machine. With the fluorescent reporter probe method, different floresence of the reporter allows for the reaction to be multiplexed.

Multiplexed Polymerase Chain Reaction

Multiplexed PCR is a methodology where several primer pairs are used in a PCR. The multiple pairs of primers will target multiple sequences in the template DNA and produce several different amplicons. In designing the primers, it is important that the annealing temperatures of the different primer pairs are similar but at the same time the target sequences are of different sizes. The amplicons can be run on electrophoresis to separate and identify the different amplicons. By performing a multiplexed PCR several amplicons can be identified in one reaction instead of just one amplicon in each reaction.

Reverse Transcriptase-Polymerase Chain Reaction

One PCR variation often used in oncology testing for detection of fusion gene products such as *BCR-ABL1* in chronic myelogenous leukemia patients is reverse transcriptase (RT) PCR. First, it is important to not call RT-PCR real time PCR because real time PCR is referring to qPCR, the technique described previously. It is possible to have a reverse transcriptase quantitative PCR, which would be termed RT-qPCR. In RT-PCR, the genetic information used is RNA and a RT is used to covert the RNA to complementary (c) DNA and then the cDNA is run on a PCR machine.

Usually, the starting material in PCR is DNA. This is because DNA is more stable when compared to RNA and it is useful in detecting point mutations, deletions, and insertions. The main issue with DNA is that DNA has long intronic regions that are not in the final transcribed mRNA product and are not ultimately in the translated product. All of the intronic regions are removed in the cell by splicing mechanisms before forming the final mRNA product. So, for fusion transcripts such as *BCR-ABL1*, by using RNA as starting material, the long intronic regions in the DNA fusion transcript are removed already. The long intronic regions are not important for the purpose of identification of the fusion transcript and can be difficult to run on an assay due to the long length of these regions.

HOT-Start Polymerase Chain Reaction

HOT-start PCR is a variation of PCR where a Taq polymerase with an attached antibody that inactivates the Taq polymerase is used. After set up of the PCR, an initial step of warming the PCR to 95°C will cause the antibody to disassociate from the Taq polymerase, thereby activating the Taq polymerase. Using a HOT-start PCR minimizes the nonspecific amplification of the DNA template that occurs during setup of the PCR before the Taq polymerase is activated.

Nested Polymerase Chain Reaction

Nested PCR is a technique that reduces nonspecific amplification of the DNA template. It is performed by two successive PCRs. The first reaction is performed with primers that cover the target sequence and some additional sequence flanking both ends of the target sequence. After the first reaction, a second reaction is performed on the products of the first PCR with primers that bind to the target sequence and are within the amplified sequence of the first PCR. This reduces the amount of nonspecific binding because in the second reaction, most of the amplicons of the first reaction only contain the target sequence and its surrounding sequences.

Sequencing

In PCRs, a specific region of a DNA template is amplified producing an abundant quantity of target sequence. However, a PCR does not actually identify the exact order of the nucleotides in a target sequence. It is especially important to know the sequence in order to discover novel mutations or to confirm mutations in PCR products. Sanger sequencing and pyrosequencing are the two main basic sequencing techniques used in the clinical laboratory and will be discussed in this section. A newer form of sequencing, next generation sequencing will be discussed in the next section.

Sanger Sequencing

During a Sanger sequencing run, four different reactions are performed. All four reactions have DNA polymerase, DNA that is being sequenced, all four dNTPs, and each reaction well will have one of the four dideoxynucleotides (ddNTPs). ddNTPs lacks a 3′-OH group that the dNTPs have. The 3′-OH group is needed to from a phosphodiester bond needed for another nucleotide to be added onto the newly synthesized sequence; and so when a ddNTP is added on, the synthesis of the new chain terminates and hence Sanger sequencing is also called "chain terminating sequencing." After the Sanger reaction is allowed to run, varying length of the target sequence is formed due to termination of synthesis of the sequence dependent on when the ddNTPs are added. Each of the products of the four reactions is run on a different lane on a gel electrophoresis resulting in four lanes. Because the ddNPT causes the synthesis of the new strands to end at different length, the products of the Sanger will be separated based on the size (length) on the electrophoresis. So by examining the four lanes you can see the order by which the nucleotide (ddNTP) was added on and be able to map out the nucleotide in the sequence.

A modification of the traditional Sanger sequencing is dye terminator sequencing. Here, instead of using four different reactions for each of the different ddNTPs, a single reaction with four different fluorescent-labeled ddNTPs is used. After a reaction run, a capillary electrophoresis is able to separate out the different length of the target sequence and identify which ddNTPs was added at the end of each sequence based on the fluorescent wavelength the strand emits. A computer is then able to translate the fluorescent signal to peaks in a chromatogram resulting in a Sanger tracing.

Sanger sequencing has been the gold standard in sequencing; however, there are certain limitation with Sanger sequencing. The first limitation is that it can only sequence a relatively short segment of target sequence at a time (800–1000 nucleotides). Another challenge with Sanger sequencing is that it is not a very sensitive test, only has a sensitivity of about 20%–25%. While this is usually not a problem for detection of constitutional mutations, this is problematic for solid tumor samples where the tumor percent is very low because in these cases a negative result can likely be a false negative result. Due to this, in molecular oncology, every clinical laboratory has cutoffs for what minimum tumor percentage is needed in the solid tumor sample for the sample to be run in an effort to decrease the chance of false negative results.

Pyrosequencing

Pyrosequencing uses a different technology from Sanger sequencing. Instead of using a "chain termination sequencing" methodology, it uses a "sequencing by synthesis" methodology. Here, one of the four nucleotides is added into the reaction well at a time and if it is complementary to the next available nucleotide on the template strand, the nucleotide will attach to the complementary nucleotide on the strand and a light signal is emitted. This light signal is captured by the pyrosequencing machine and plotted on a pyrogram. When the complementary nucleotide attaches to the strand, a pyrophosphate (PPi) is released. The PPi is converted to ATP with ATP sulfurylase and the ATP then helps in the conversion of luciferin to oxyluciferin that emits a light signal dependent on the quantity of ATP. When many repeats of the same nucleotide are present on a strand (e.g., AAAAAA), the signal is stronger and has a larger peak on the pyrosequencing tracing. If the nucleotide is not complementary, the nucleotide will not attach to the strand and no light signal is emitted. The residual complementary nucleotide that does not attach to the strand and the noncomplementary nucleotide that does not attach to the strand will be degraded with apyrase. Once the residual nucleotides are degraded, another one of the four nucleotides will be added and this process will begin again. Once all four nucleotides have been added, this cycle will repeat until the target sequence is completely sequenced. So, in sum, the nucleotides are added one at a time, and based on whether a light signal is emitted from the reaction, the pyrosequencing machine is able to know whether a nucleotide is present or absent at that location. Pyrosequencing is a highly accurate method of sequencing; however, the main challenge here is that it can only sequence 300–500 nucleotides at a time.

Next Generation Sequencing

Next generation sequencing has garnered much attention in recent years. Conceptually, the main difference between next generation sequencing and more traditional sequencing methods such as Sanger sequencing is that next generation sequencing is able to sequence a lot more nucleotides in a single run, a concept we refer to as high throughput. Instead of the 800–1000 nucleotides sequenced during a typical Sanger sequencing run, the next generation sequencing platforms can sequence millions of nucleotides during a run. Another major difference is that next generation sequencing is massively parallel; and so we can multiplex many reaction and multiplex different specimens all on one run. The last major difference is that next generation sequencing can perform ultra-deep sequencing, allowing the sequencing to be highly sensitive, unlike the relatively insensitive sequencing of Sanger sequencing (approximately only 25% sensitivity). These advantages to traditional sequencing allow for sequencing of more genes and more targets at a faster pace and a lower cost. This has all become quite useful in the drug discovery process and in clinical testing.

Sequencing can be thought of in terms of whole genome sequencing, whole exome sequencing, full gene panels, and hot spot panels. In whole genome sequencing, the entire patient's genome is sequenced. In whole exome sequencing, the whole exome of the patient is sequenced. In full gene panels, entire target gene/s are sequenced; and in hot spot panels, only certain portions of a gene/s are sequenced. The amount and where we sequence is dependent on the purpose of performing the sequence. For example, in a pediatric patient with an unknown or uncharacterized syndrome, whole genome sequencing can be performed to look for genetic aberrations in their genome that might be correlated with their syndrome. On the other hand, for a pediatric oncology patients being evaluated for enrollment into clinical trials, their whole exome might be sequenced to match them with a suitable clinical trial based on the mutations they harbor. However, for an adult patient with thyroid cancer, a targeted hot spot panel might be employed to test for the common mutations seen in patients with thyroid cancer.

Many different platforms and technologis are available for next generation sequencing; however, the basic concept for all the platforms is the same. In next generation sequencing, there are seven major steps. The first step is library preparation (1), where the target sequences are fragmented and the fragments are modified with adapters and barcodes to identify the sequence and to act as reference points during the next steps. Next, during clonal amplification (2), the individual sequences are clustered/compartmentalized allowing each individual sequence to be amplified many folds in a cluster/compartment. In the third step, sequencing (3), these individual sequences that are clustered/compartmentalized and amplified is sequenced. Then, the sequences are mapped and aligned (4) to reference sequences by the information's system. Next, variants are called (5) based on parameters set by the molecular team during validation of the specific assay. Then, these variants are annotated (6) based on available literature and databases. Lastly, a molecular pathologist will interpret the variants (7) and issue a report taking into account the patient's clinical history.

While multiple different platforms for next generation sequencing are available in the clinical laboratory, currently there are two main technologies used. The first is the Illumnia platform and the second is the Ion Torrent platform, both of which will be discuss here.

Illumina Platform

The Illumina platform "sequences by synthesis" uses a dye sequencing and camera capture technology. After the DNA is purified, the DNA is fragmented and adapters and barcodes (need to be used in multiplexing samples) are added. These modified DNA fragments are placed into a flow cell where oligonucleotides that are complementary to the adapters, which were added to the sample DNA fragments, are attached to the flow cells. The adapters attach to the complementary oligonucleotides and unattached fragments are washed away. Through a technology called bridge amplification, each of the attached DNA strands will amplify and create clusters of approximately 1000 identical strands. Then, primers and all four modified nucleotides (modified with 3′ reversal blockers and labeled with fluorescence, each of the four nucleotides will have a different fluorescent color) will be added to the reaction. During the first cycle, the primer will attach and then the complementary nucleotide to the next available nucleotide will attach to the 3′ end of the primer. The rest of the unattached nucleotides will then be washed away. Each of the clusters will then have a fluorescent color based upon the nucleotide that was added. A camera will then take an image of the flow cell, and each different fluorescence signal will represent the specific nucleotide (A, T, C, or G) that was added to all the individual single strands of an individual cluster (~1000 strands). Next, a chemical is used to remove the 3′ reversible blocker on the newly attached nucleotide, and then the flow cell is again flooded with the four modified nucleotides. This process is repeated until the whole new complementary strand is synthesized. The sequence of each of these clusters can be identified based on the order of the fluorescence emitted by the each of the clusters. So, in essence, there are thousands of sequencing reaction being performed in one assay run (each cluster is a sequence), and this is why next generation sequencing is also referred to as "massively parallel sequencing." Once the newly synthesized strand is synthesized, it is washed away and from the original forward strand and a reverse strand is synthesized. The reverse strand is then sequenced based on the same methodology as the sequencing of the forward strand. Sequencing the reverse strand is called "pair-end sequencing" and is unique to the Illumnia technology. This has an advantage of helping detect gene fusions and translocations in RNA sequencing.

Currently, Illumina has a few platforms that utilize this technology. The main difference in the platforms is the throughput (how much DNA can be sequenced in one run). In increasing order of throughput: MiSeq Series (15 gb max output, max read number 25 M, max read length 2 × 300 bp); Next Seq Series (120 gb max output, max read number 400 M, max read length 2 × 150 bp); HiSeq Series (1500 gb max output, max read number 5000 M, max read length 2 × 150 bp); HiSeq X Series (1800 gb max output, max read number 6000 M, max read length 2 × 150 bp); and HiSeq extended when you put many Hiseq together. The specific platform chosen is based on the clinical application of the platform. If a large panel or multiplexing many different samples is needed, then a platform with higher throughput should be chosen; and if only a few genes and/or a few samples are tested, a platform with lower throughput should be chosen. Another terminology that is important to understand for next generation sequencing is depth of coverage. Depth of coverage refers to how many clusters had sequenced that particular nucleotide or segment of sequence. The deeper depth of coverage means that the nucleotide/s in questions have been sequenced more times, and the molecular pathologist interpreting the sequences can be more confident of the presence or the absence of a mutation.

Ion Torrent Platform

The Ion Torrent platform also uses a "sequencing by synthesis" methodology. The main difference of this platform from other next generation sequencing techniques is that instead of using optics to detect the incorporation of nucleotides, it detects nucleotide incorporation by monitoring changes in the pH. Conceptually, the major steps are very similar to the Illumina technology. The first main difference is that instead of using bridge amplification to clonally amplify the individual fragments of DNA, it uses a methodology termed emulsion PCR. Here, each individual strand is compartmentalized into an emulsion bead where PCR occurs to clonally amplify fragments of the target sequence. This essentially allows thousands of individual PCRs to occur simultaneously. Once the segments of sequences are clonally amplified, it is loaded onto a computer chip where each emulsion bead will be loaded into a microwell on the chip. For the sequencing step, one of the four nucleotides is flooded into the microwells. If the nucleotide is complementary to the template strand, it will incorporate onto the strand and a hydrogen ion will be released causing a pH change that can be detected by the Ion Torrent sensors. If it is repeats of the same nucleotide on the template strand, multiple nucleotides will be incorporated and multiple hydrogen ions will be released causing a higher pH change that can be detected by the sensors in the microwell. Next, the nucleotides are washed away, and another nucleotide is added and the same process is repeated until all four nucleotides are added. This cycle repeats until the template sequence is completely sequenced. Then, the raw data is processed and analyzed similarly to the Illumina platform.

Ion Torrent also has several platforms using their technology. The difference in the platforms is throughput and automation. The machine with the lowest throughput is the personal genome machine, the next is the proton, and now there is the S5 XL, which is more flexible in that it can run lower throughput runs as well as higher throughput runs and it is more automated. Three commonly used chips with different capacity include the 314 (1.2 million wells, 1 sample), 316 (6 million wells, 2–7 samples), and 318 (12 million wells, 8–10 samples). One of the main advantages of the Ion Torrent technology when compared to the Illumnia technology is that the DNA quantity requirements are much lower. This makes the Ion Torrent technology ideal for solid tumor specimens where available tumor DNA is usually less than in hematologic malignancy and testing for constitutional defects. One of the main disadvantages is that when the repeats are long, due to the technology used to detect the sequence, it may not always be as accurate in quantifying the number of repeats in the sequence.

KEY POINTS

- A DNA nucleotide is composed of a deoxyribose (five carbon sugar) with an attached nitrogenous base (adenine [A], guanine [G], cytosine [C], or thymine [T]) and a phosphate group.
- In contrast to a DNA nucleotide, an RNA nucleotide has a ribose backbone instead of a deoxyribose backbone; and while they have the same nitrogenous bases of guanine, adenine, and cytosine, instead of thymine, RNA nucleotides have uracil.
- Two polynucleotide strands are attached to each other by hydrogen bonds: Adenine will only bind to a thymine (two hydrogen bonds) and cytosine will only bind to a guanine on the other strand (three hydrogen bonds) forming a DNA double helix.
- The double helix DNA is wrapped around histone proteins, and these proteins are packed tightly together until they form a chromosome.
- Replication occurs when DNA strands are copied resulting in identical copies of the original (template) DNA and occurs in three steps three steps: Initiation, elongation, and termination.
- The central dogma in molecular genetics is that DNA is transcribed into RNA which is translated into proteins.
- Nomenclature can be a subject of confusion in molecular diagnostics and three nomenclatures commonly used in molecular diagnostics are coordinates of a nucleotide, cytogenetic nomenclature, and mutation nomenclature.
- In general, a cytogenetic technique such as conventional karyotyping, virtual karyotyping, or FISH gives an overview of the entire DNA genome; PCR techniques give information on specific parts of the DNA genome; and sequencing techniques gives a detailed of view the DNA genome.
- It is important to keep in mind minimal amount of DNA, tumor heterogeneity, and how the tissue is stored as these factors can affect testing results.
- Cytogenetic karyotyping is a cytogenetic technique that allows for an overview of cytogenetic abnormalities such as aneuploidy (changes in chromosome numbers), translocations, inversions, and deletions of the patient's chromosomes.
- FISH is a cytogenetic technique that uses fluorescent-labeled probes to detect the absence or presence of specific regions/sequences on chromosomes.

- Two main types of virtual karyotype, aCGH and SNP array are used.
- The basic goal of PCR is to amplify many copies of a target DNA in order to examine that particular focus of DNA more closely and this is done through multiple thermal cycles that consist of three steps: Denaturing, annealing, and extension.
- When designing a PCR, three main things needs to be considered: Primer design, reagent concentrations, and thermal cycling conditions.
- Some common variations of PCR include qPCR, multiplexed PCR, RT-PCR, HOT-start PCR, and nested PCR.
- Sequencing identifies the exact order of the nucleotides in a target sequences and the two most common first generation sequencing technologies includes Sanger sequencing and pyrosequencing.
- Next generation sequencing refers to sequencing technologies that have very high throughput, meaning that they are able to sequence a lot more nucleotides at once.
- The Illumnia and Ion Torrent platforms are the two main next generation sequencing platforms used clinically; however, regardless of platform used in next generation sequencing, they all consists of seven major steps.

REFERENCES

[1] Miller MB, Tang YW. Basic concepts of microarray and potential applications in clinical microbiology. Clin Microbiol Rev 2009;22:611–33.

[2] Zheng L, Shen B. Okazaki fragment mutation: nucleases take center stage. J Mol Cell Biol 2011;3:23–30.

[3] Collins FS, Morgan M, Patrinos A. The human genome project: lesson from large scale biology. Science 2003;300:286–90.

[4] Rosenbloom KR, Armstrong J, Barber GP, Casper J, et al. The UCSG genome browser database: 2015 update. Nucleic Acids Res 2015;43(Database issue):D670–81.

[5] Simons A, Shaffer LG, Hastings RJ. Cytogenetic nomenclature: changes in the ISCN 2013 compared to 2009 edition. Cytogenet Genome Res 2013;141:1–6.

CHAPTER 13

Application of Molecular Diagnostics

Contents

INTRODUCTION

In Chapter 12, Overview of Molecular Diagnostics Principles, basic molecular diagnostics techniques were introduced. Here, the clinical application of these techniques in the fields of molecular oncology, inherited disease, infectious disease, and pharmacogenomics will be presented. This section will highlight some major applications of molecular diagnostics

Microbiology and Molecular Diagnosis in Pathology.
DOI: http://dx.doi.org/10.1016/B978-0-12-805351-5.00013-2

that will help in the understanding of how molecular techniques are used in clinical practice; however, this is by no means a comprehensive or exhaustive list of all the applications in clinical practice.

MOLECULAR ONCOLOGY

Cancers that occur within a given tissue have distinct genomic changes which define each individual tumor in terms of molecular processes involved in the development of tumor [1]. Molecular oncology can be divided into two categories, hematologic malignancies and solid tumors. Traditionally, much more research has been done in hematologic malignancies when compared to solid tumors because of the accessibility, relative homogeneity, and liquid nature of certain hematologic malignancies. However, in the past few years, more research is being performed in solid tumors and it is changing the way surgical pathology is being practiced. In the following sections, solid tumors will be discussed first, followed by a discussion on hematologic malignancies.

Molecular diagnostics aids in defining many aspects of human neoplasm. It can help in defining the diagnosis, prognosis, and treatment; and it can be used to monitor the disease. Recently, a lot of excitement has surrounded molecular diagnostics in the oncology field due to the ability to guide treatment with molecular tests. A popular term that has been used to describe this is companion diagnostics. Companion diagnostics means pairing a diagnostics test with a drug to predict how well the drug will work on the patient before giving the drug to the patient. This is especially important in the field of human oncology due to the high cost and the many toxic side effects of oncological drugs. The basic premise behind companion diagnostics is that it is better to spend a few hundred dollars to run a diagnostics test on a patient that can possibly spare a patient from going through an oncology treatment regimen that cost thousands of dollars and causes severe and toxic side effects when the drug does not even work on the patient. A common example of companion diagnostics is chronic myelogenous leukemia (CML), where it was discovered that Imatinib works extremely well on the patients with the *BCR-ABL1* translocations, which has become the defining criteria of CML.

One important concept to understand about in molecular oncology is Knudson's hypothesis [2]. The general idea in this hypothesis is that multiple "hits" (mutations) are needed for oncogenesis. Specifically, this hypothesis describes a "two hit" theory, where the patient has a mutation inherited in his/her genome, and when a second mutation is acquired, the patient rapidly develops a malignancy. The support for this hypothesis comes from Knudson's work on retinoblastoma. This concept in tumor-genesis is also termed as "loss of heterozygosity," where a patient is originally heterozygote for a gene by inheriting a mutated copy/allele of a gene (this is the "first hit" and is usually a point mutation). Then, a "second hit" occurs (usually a deletion, point mutation, or methylation) of the second unmutated copy/allele of the gene. This "second hit" is an acquired mutation and causes the patient from being heterozygote for the gene to becoming having both genes mutated; and thereby "loss of heterozygosity" occurs. Knudson's hypothesis is based on tumor suppressor genes, which are genes that suppress the uncontrolled growth of cells. If a tumor suppressor gene is mutated, rendering it not as effective in controlling cell growth, a tumor develops. Some very common tumor suppressor genes include *TP53* located on 17q13, *RB* located on 13q14.1, *APC* located on 5q23, and *VHL* located on 3p26-25. Understanding that in "loss of heterozygosity" testing the patients are essentially having a bi-allelic mutation is important in that the allele needs to be separated in this testing. To identify alleles, surrogate markers such as short tandem repeats (short sequences that repeat for a variable number of times) or single nucleotide polymorphisms (SNPs) are identified in polymerase chain reaction (PCR).

While Knudson's hypothesis is focused on the deactivation of tumor suppressor genes, it was later found that tumor-genesis also occurred by the activation of proto-oncogenes. Oncogenes encode proteins that affect cell growth, differentiation, apoptosis, and signal transduction. Here, the activation of a gene with a point mutation, amplification, or translocation is needed instead of a deletion of the gene for the patient to experience tumor-genesis. In oncogenes, only one copy needs to be mutated for a tumor to develop and examples of common oncogenes include *K-RAS* on 12p12, *MYC* on 8q24, and *BRAF* on 7q34.

Solid Tumor

In recent years, molecular diagnostics has permeated into most solid tumor testing; however, only some of the specific testings are highlighted in this section. The specific testing that is highlighted is not selected based on importance, but rather on their ability to exemplify how different molecular techniques can be utilized on these diseases.

Besides the specific mutations that are usually tested for in solid tumors, many institutions have started utilizing a strategy that uses large next-generation sequencing panels to test many genes at once. Additionally, it is now possible to detect fusion genes through RNA next-generation sequencing and even detect amplifications of genes with next-generation sequencing techniques further adding to the information that next-generation sequencing can yield. In some cancer centers in the United States, laboratory scientists are performing next-generation gene panels (from a few genes to hundreds of genes) on all patients with malignancy. Also, many commercial laboratories have developed next-generation sequencing assays that also have a few hundred genes and they serve as reference laboratories for smaller laboratories that cannot support their own next-generation sequencing laboratories. This is done to help in the selection of patients for clinical trials and also to look for new possible biomarkers for different tumors while concurrently testing for the specific genes that are currently clinically actionable in these patients. This practice is also producing large amounts of data that can be used in future research. With such a large amount of data being produced with next-generation sequencing assays, it is sometimes hard to make sense of what is clinically significant and what is not. The mutations that are found are compared to databases such as COSMIC and a review of the literature is done on mutations not often seen to determine the significance of the mutations.

A subject worthy of discussion is the distinction between germline versus somatic mutations. A germline mutation is a mutation that is present in an egg or a sperm, is heritable, affects all cells in an offspring, and can cause a cancer family syndrome (e.g., Li Fraumeni, hereditary retinoblastoma, Cowden, hereditary breast/ovarian cancer, Lynch syndrome, and Von Hippel-Lindau syndrome). On the other hand, a somatic mutation is a nonheritable mutation that occurs in nongermline tissues such as breast, colon, and thyroid. When a next-generation sequencing assay is ran on a solid tumor sample alone it is impossible to tell whether the mutation detected is due to a germline mutation or a somatic mutation. Some mistakenly try to infer the origin of the mutation based on the relative amount of that mutation to unmutated sequences. The idea is that if a mutation has a very high quantity relative to wild type in a sample, it is most likely due to a germline mutation and vice versa; however, this is not reliable because of the heterogeneity of a solid tumor specimen. A more reliable method that can be used to help discriminate between germline and nongermline mutations is to run a solid tumor sample along with a normal sample from the patient (e.g., blood) simultaneously. This way, if the mutation is only present in the tumor sample and not in the normal sample, it is most likely a somatic mutation and if it is present in both the tumor sample and germline sample, then it likely germline. The reason why one might want to determine whether a tumor is germline versus somatic is that if it was germline, family members of the patient can obtain genetic counseling and possibly get tested for the mutation. However, this brings up a controversial topic of whether consent needs to be obtained to perform germline testing on oncology specimens. In general, for any constitutional genetics testing, consent needs to be obtained because the patient might not want to know if they have a germline mutation and neither does their family that might be affected by such a mutation [3].

Colorectal Tumors

One area of interest in molecular testing of colorectal tumors is microsatellite instability (MSI) testing. While MSI does not have specific therapeutic targets, it has certain prognostics implications. In patients with MSI, standard chemotherapy treatment such as 5-fluorouracil is not as effective and also patients with MSI has improved survival. MSI testing is also used to help detect Lynch syndrome (an autosomal dominant inherited syndrome also known as hereditary nonpolyposis colorectal cancer, HNPCC), where germline mutations in mismatch repair genes *MLH1*, *MSH2*, *MSH6*, and/or *PMS2* are present. MSI can also be seen in sporadic colorectal cancer through acquired hypermethylation of *MLH1*.

MSI occurs when there is expansion or contraction of microsatellite repeats. Microsatellite repeats occur in the noncoding region of DNA and consists of short (one to six) nucleotide DNA sequences that repeat multiple times. For example, a mononucleotide repeat is (AAAAAAA) and a di-nucleotide repeat is (ATATATATAT). These microsatellite repeats are supposed to only repeat a set number of times; however, slippage can occur during DNA replications, and that can cause the microsatellite repeats to repeat a different number of times. When this happens, it is usually fixed by the mismatch repair genes; but when there is a defect in mismatch repair genes, the abnormal number of repeats is not fixed and MSI is present.

Many different techniques can be used to test for MSI. The three main techniques are PCR; sequencing; and interestingly, a technique that is not molecular, immunohistochemistry (IHC). With PCR, the regions that are likely to harbor MSI are amplified (two mononucleotide repeat markers: BAT25 and BAT26; and three di-nucleotide repeat markers: D2S123, D5S346, and D17S250 are usually used) and the PCR products are run on a capillary electrophoresis to determine whether different size repeats (representing MSI) are present. In sequencing, the mismatch repair genes are sequenced to identify any mutations. Lastly, with IHC, mismatch repair proteins are stained as a surrogate test for mutations in mismatch repair genes. In IHC, the lack of the stain, or lack of mismatch repair (MMR) expression, is positive for MSI. It is important to note that *MLH1* complexes with *PMS2* and *MSH2* complexes with *MSH6*. Also, understanding that *PMS2* needs to complex to *MLH1* and *MSH6* needs to complex with *MSH2* while the reverse is not true is important. This lets us understand why when *MLH1* is negative *PMS2* is usually negative and when *MSH2* is negative *MSH6* is usually negative, and why the reverse is not necessarily true.

IHC and PCR can be thought of complementary testing to help identify MSI. Studies have shown that when used together they increase the sensitivity for the detection of MSI. However, IHC is an easier and cheaper test to perform and is a good surrogate marker for MSI. In this age of cost containment in health care, it will be interesting to see how each laboratory/institution formulates testing algorithms for their site. While many different algorithms to test for MSI (not just identification of MSI but also to determine if it is due to Lynch syndrome or a sporadic mutation) is available, one common testing algorithm that is used will be presented here. First, IHC is performed on the *MLH1*, *PMS2*, *MSH2*, and *MSH6*. If all are negative, then there is likely no MSI and further testing by PCR will only be performed if clinical suspicion is high. If any of these stains are positive, there is MSI; however, additional testing is performed to identify Lynch syndrome versus sporadic mutation. If *MLH1* and *PMS2* IHC is positive, *BRAF* V600E testing is performed and if positive, the mutation is most likely sporadic. However, if *BRAF* V600E is negative, performing a promoter methylation study or gene sequencing is recommended. If the initial IHC is positive for *MSH2* and *MSH6*, *MSH6*, or *PMS2*, then sequencing studies are recommended to determine the origin of the mutation. So, in this algorithm, the first screening test for this mutation is not a molecular test; but rather, it is a simple IHC.

Another important aspect of colorectal cancer molecular testing is testing *KRAS2* in codon 12, 13, 61, and sometimes 146 [4]. Testing *KRAS2* is important in colorectal cancer testing because in patients with stage IV colon cancer with this mutation, *EGFR* inhibitor therapy (e.g., Cetuximab) is not as effective. This is an example of using molecular testing as a companion diagnostics test for oncology drugs.

This discussion of molecular testing in colorectal cancers would not be complete without discussing familial adenomatous polyposis (FAP) [5]. In patients with FAP, they have an *APC* knockout mutation and clinically they have multiple adenomatous polyps that have a 100% risk of transforming into colorectal cancer if the colon is not removed. The average age of onset for these patients is 39 years old. Multiple syndromes are associated with FAP including Gardner's syndrome (FAP, desmoids tumors, osteomas, epidermoid cysts, and dental abnormalities) and Turcot's syndrome (FAP with brain tumors).

Breast Tumors

A few different types of molecular assays are available for breast tumor testing. First, commercial gene panels are available for prognosis and/or to help guide treatment decisions. The two most famous gene panels are OncotypeDx and Mammaprint. OncotypeDx helps guide treatment in female patients by predicting the benefit of chemotherapy and the likelihood of distant breast cancer recurrence in patients who are *HER2* negative, estrogen receptor positive, and have invasive breast cancer. Mammaprint helps predict the chance of breast tumor metastasis in women under 61 that have lymph node negative breast cancer (can be estrogen positive or estrogen negative).

Another well-documented test in breast cancer is the detection of *HER2* amplification. It is important to identify these patients because a monoclonal antibody (Herceptin) against the *HER2/neu* receptor works well in breast cancer patients with *HER2* amplification. An algorithm used for *HER2* detection is that all newly diagnosed breast cancer patients and patients who develop metastatic disease should be tested for *HER2* amplification by *HER2* IHC stain. The stain will be graded from a scale of 0 to 3+ (0 [negative], 1+ [negative], 2+ [equivocal], and 3+ [positive]); and if the stain is 2+, the IHC will be followed up with fluorescence in situ hybridization (FISH) testing for *HER2* amplification. As in MSI testing, IHC instead of a molecular test is usually used first in many testing algorithms.

Another often tested mutation is in hereditary breast cancer. Here, the two genes that are tested for are the *BRCA1* (17q) and *BRCA2* (13q), both of which are tumor suppressor genes involved in DNA repair. Ten percent of breast cancer and fifteen percent of ovarian cancer is caused by *BRCA1* and *BRACA2*. With *BRCA1*, there is a 65%–80% increased lifetime risk for breast cancer and a 40%–50% risk for ovarian cancer. Similarly, with *BCRA2*, the lifetime breast cancer risk is 45%–85% and it is 10%–20% for ovarian cancer. *BRAC1* and *BRAC2* mutations are tested by a germline sequencing based test, meaning that the sample is from a constitutional source such as whole blood. Since it is a germline test, genetic counseling (both pre- and posttest) is required for the patient [6].

Thyroid

In organs such as the thyroid, molecular diagnostics can aid it in the identification of certain tumors. In papillary carcinoma, a common mutation is an activating mutation, *BRAF* V600E, on exon 15. Besides the conventional papillary carcinoma, this mutation is seen in 70% of tall cell papillary carcinoma and is also seen in other cancers such as colon cancer and melanoma. When used in final needle aspiration specimens, studies have shown that identification of this mutation adds to the sensitivity in suspicious lesions. Other mutations seen in papillary carcinoma include the *RET/PTC* translocation and *RAS* mutations. In follicular carcinomas, *RAS* mutations are also seen along with *PPARγ/PAX8* translocations.

Brain Tumors

Brain tumors are another area with more molecular biomarkers being discovered. In grade 2/3 oligodendroglioma codeletion of 1p/19q shows improved response to chemo-radiation and increase survival. Traditionally, this has been tested by FISH and by loss of heterozygosity testing with PCR; however, this is a mutation that can now be tested by newer methods such as SNP arrays. In another mutation of interest, *IDH1* R132H and *IDH2* R172K is mutated in 80% of grades 1 and 2 astrocytomas and oligodendrogliomas and >90% glioblastoma multiforme. Presence of these two mutations suggests an infiltrative glioma with longer survival. These mutations can be tested with either a hotspot PCR or antibody testing. Also, in glioblastoma multiforme, *MGMT* helps in resistance against alkylating agents; however, if the promoter region of *MGMT* is methylated, *MGMT* can no longer help in the resistance against alkylating agents and portends an improved survival for these patients. Lastly, *EGFR* amplification is seen in 40% of glioblastoma multiforme and if *EGFR* is seen in a grade 2 astrocytoma, it likely indicates that the lesion is a higher grade and the lower grade was originally diagnosed due to poor sampling.

Lung Cancer

Molecular diagnostics has changed lung cancer in many meaningful ways, especially in terms of treatment. Various mutations have been shown to affect the response to therapy of lung cancer patients. These genes usually affect the *RAS-RAF-MAP* and *PI3K-AKT* pathways in the cells and these genes are important in that many of these genes have targetable therapy.

ErbB1, more commonly known as *EGFR* and located on 7p12, is a good example of how testing of these genes have become commonplace in a subset of patients. This gene codes for a trans-membrane receptor tyrosine kinase that has been shown to signal through the aforementioned pathways. Mutations occur in nonsmokers, females, adenocarcinomas, and East-Asians commonly on exon 18, 19, 20, and 21 with exons 19 and 21 comprising 90% of the mutations found in these patients. Molecular assays that test for these mutations are a good example of a companion diagnostic test because drugs that work well on patients with these mutations are currently available. For example, *EGFR* targeted therapy such as monoclonal antibodies that inhibit the extracellular domain of *EGFR* (Cetuximab and Panitumumab) and tyrosine kinase inhibitors (Erlotinib and Gefitinib) are available.

EML4-ALK translocation, like *EGFR* mutations, occurs more commonly in nonsmokers. This is in contrast with *KRAS* mutations which occur more commonly in smokers. *EML4-ALK* also occurs more commonly in adenocarcinomas and is an important mutation to test for because treatments such as Crizotinib that targets the chimeric *ALK* tyrosine kinase are available.

Besides using mutations to select therapy, resistance to therapy by mutation testing is also available. They include *EGFR* mutations exon 20 T790M and exon 19 D761Y, *MET* amplification and point mutations (20% tumor resistant to *EGFR* therapy), *RAS* mutations (*KRAS*, *NRAS*, *HRAS*; however, almost are *KRAS*), and *BRAF* mutation (activating mutation on exons 11 and 15 with mutations in V600E accounting for 40%).

Since *BRAF* mutations have been mentioned many times already, it is worthwhile to briefly discuss *BRAF* here. The *BRAF* gene makes the protein B-Raf that is involved in cell growth. *BRAF* mutations can be seen in inherited birth defects and can cause cancer. This is a mutation that can be seen in many cancers including malignant melanoma, lung cancers, lymphomas, and even hairy cell leukemia. The *BRAF* V600E mutation is especially famous due to the good response rate of Vemurafenib

(B-Raf enzyme inhibitor) to unresectable or metastatic melanoma patients with this specific mutation.

Translocations in Sarcomas

As in hematologic malignancies, many translocations are seen in sarcomas. A list of common translocations in sarcomas is listed in Table 13.1. In sarcoma, many of the translocations are an objective method to diagnose the sarcoma; however, these translocations may also have prognostic and therapeutic implications. Many different techniques can be used to identify these translocations. For example, in Ewing's sarcoma, a *EWS* FISH break-apart probe can be a sensitive screening test to determine whether there is a translocation on *EWS*. Here, a normal patient usually has two yellow signals in a cell. In a patient where the *EWS* has translocated, there is a green and red signal (representing the gene translocation) and a yellow signal (intact copy). In certain protocols at certain institutions, >5%

Table 13.1 Common translocations in sarcomas

Tumor name	Genes	Translocation
Ewing's sarcoma/peripheral primitive neuroectodermal tumor	*EWSR1/FLI1*	t(11;22)
	EWSR1/ERG	t(21;22)
	EWSR1/ETV1	t(7;22)
	EWSR1/FEV	t(17;22)
	EWSR1/E1AF	t(2;22)
Clear cell sarcoma	*EWSR1/ATF1*	t(12;22)
Desmoplastic small round cell tumor	*EWSR1/WT1*	t(11;22)
Alveolar soft part sarcoma	*ASPSCR1/TFE3*	t(X;17)
Dermatofibrosarcoma protuberans	*COL1A1/PDGFB*	t(17;22)
Synovial sarcoma	*SYT/SSX(1)(2)(3)*	t(X;18)
Myxoid liposarcoma	*TLS/CHOP*	t(12;16)
	EWS/CHOP	t(12;22)
Alveolar rhabdomyosarcoma	*PAX3/FKHR*	t(2;13)
	PAX7/FKHR	t(1;13)
Congenital fibrosarcoma	*ETV6/NTRK3*	t(12;15)
Angiomatoid fibrous histiocytoma	*FUS/ATF1*	t(12;16)
Endometrial stromal sarcoma	*JAZF1/JJAZ1*	t(7;17)
Low grade fibromyxoid sarcoma	*FUS/CREB3I2*	t(7;16)
Inflammatory myofibroblastic tumor	*TPM3/ALK*	t(1;2)
	TPM4/ALK	t(2;19)
Myxoid chondrosarcoma	*EWS/CHN*	t(9;22)
	TFC12/CHN	t(9;15)
	TAF2N/CHN	t(9;17)

break-apart of the probe is abnormal. If the FISH is positive or if there is high clinical suspicion of a Ewing's sarcoma, a more specific test like reverse transcriptase polymerase chain recation (RT-PCR) can be performed to identify the mutation. RT-PCR can be run on the sample with primers testing for the two most common translocations in Ewing's sarcoma, the *EWSR1/FLI1* t(11;22) translocation and the *EWSR1/ERG* t(21;22) translocation. If the RT-PCR is positive, Sanger sequencing or other sequencing techniques can be used to confirm the translocation.

APPLICATION OF MOLECULAR DIAGNOSIS IN HEMATOLOGY

Historically, molecular hematopathology has grown faster than molecular solid tumor as a field mainly because the samples are easier to work with in the molecular laboratory. This is because it is easier to obtain a fresh sample (from a blood draw or bone marrow aspirate) and the liquid nature of "wet" hematopathology allows for the cells to come already disassociated as opposed to solid tumors where the cells are not (exception being "dry" hematopathology (lymphomas)). While most of the solid tumor diagnosis is largely based on morphology; in hematopathology, many diagnosis and classification schemes in the WHO for hematopathology is based on the presence of specific molecular alterations.

For solid tumors, the current practice algorithm usually consists of diagnosis based on H&E morphology. This might be complemented by IHC staining and a molecular testing in certain circumstances as discussed above. However, in the current hematopathology practice algorithm it is usually a lot more complicated. H&E morphology; immunophenotype by flow cytometry and IHC; cytogenetic karyotyping and FISH analysis; mutation testing/specific fusion transcript detection by PCR, fragment analysis; or sequencing is usually all part of the work up for a hematopathology malignancy. Now, with the advent of next-generation sequencing panels in the clinical laboratories, these panels are also slowly being incorporated into daily practice. Furthermore, unlike in solid tumor testing, where most of the time, the sample comes from paraffin embedded tissue; in "wet" hematopathology the specimen is usually collected fresh and multiple media can be used. One general rule to keep in mind is that bone marrow and leukemic blood cells should be collected in general abbreviation no need to expand (EDTA) (purple top tubes) for molecular testing and samples for cytogenetic testing should be collected in sodium heparin (green top tubes).

One important type of testing used in hematopathology testing that is not readily used in solid tumor is clonality testing. In humans, there are two main types of immunity cells, B-cells and T-cells. During an immune response their receptors are able to rearrange in many different combinations in order to identify different antigens (present on pathogens) in the body and to be able to eliminate them. On the B-cell immunoglobulin locus there is the heavy chain (IGH), kappa light chain, and lambda light chain. In the IGH locus, there are variable diversity segments, joining segments, and constant segments; and different combinations of these segments creates the diversity that is needed to be able to identify the vast amounts of antigens that can be present. The T-cell receptor locus consists of gamma, beta, delta, and alpha, which also like the IGH can rearrange to have the diversity needed on the T-cell receptor to recognize the vast amounts of antigens. So, a specific T-cell or B-cell clone (specific rearrangement) may be present in a great amount due to that specific clone being preferentially proliferated in a hematopathologic malignancy.

The main indication to perform T-cell or B-cell clonality testing is to aid in the distinction of a benign versus malignant neoplasm in an atypical lymphoid proliferation. While flow cytometry can be used to determine clonality (kappa vs lambda) in B-cells, it is not useful for T-cells and also flow cytometry is limited in that intact cells are needed and it cannot be performed on formalin-fixed paraffin-embedded (FFPE) tissue. On the other hand, T-cell and B-cell clonality testing is limited by the fact that a clone does not equal malignancy (it can also be due to a reactive condition such as lymphomatoid papulosis); and also if no clones are detected by an assay, it does not mean that there is no malignancy because the assay does not test for all the possible clones. T/B-cell gene rearrangement studies are basically performed by multiplex PCR reactions. Primers are designed to only amplify when they are rearranged since they are very far apart before they are rearranged. Standardization has occurred with the primers for T-cell and B-cell rearrangements where most clinical laboratories use the primer sequences from the BIOMED-2 guidelines from Europe. After the PCR is performed, the products are run on a gel or capillary electrophoresis. In general, a gel that has a smear is polyclonal and if bands are present, it is monoclonal. On capillary electrophoresis, peaks that form a Gaussian distribution are polyclonal and large peaks that are not part of the distribution are monoclonal. Some common peaks seen include stutter peaks and bi-allelic peaks.

Acute Leukemia

The classification of acute leukemia has changed from the French–American–British classification which is largely based on morphology to the WHO classification where many entities are defined by cytogenetic abnormalities.

In the WHO classification, a portion of the acute myeloid leukemia (AML) is based on these recurrent cytogenetic abnormalities. The most significant are three specific abnormalities in AML where they are considered AML regardless of bone marrow blast count. They include AML with the translocation of *RUNX1-RUNX1T1* t(8;21)(q22;q22); AML with inversion of *CBEB-MYH11* inv(16)(p13.1q22) or t(16;16)(p13.1;q22); and the translocation of *PML-RARA* t(15;17)(q22;q12), also known as acute promyelocytic leukemia (APL). Of note, determination of APL in a timely manner is of vital importance due to the great response for patients to all-trans retinoic acid, a form of vitamin A, when given in a timely manner. Other abnormalities under the AML with recurrent cytogenetic abnormalities category include AML with t(9;11)(p22;q23), *MLLT3-MLL*; t (6;9)(p23;q34); *DEK-NUP214*; inv(3)(q21;q26.2) or t(3;3)(q21;q26.2), *RPN1-EVI1*; and AML (megakaryoblastic) with t(1;22)(p13;q13), *RBM15-MKL1*. Other significant mutations in AML include *FLT3*-ITD (internal tandem duplication) which is over expressed in 20%–30% of AML blasts and associated with reduced survival; *FLT3*-TKD with unclear clinical significance; *NPM1* mutation which is the most common AML with normal cytogenetics and has a good prognosis if *FLT3*-ITD negative; and homozygous or bi-allelic *CEBPA* mutation that has a good prognosis.

In acute lymphoblastic leukemia (ALL) patients, many cytogenetic abnormalities have prognostic implications. For example, infants with t(4;11) have a poor prognosis. Also, children with t(12;21) have a good prognosis; however, this mutation is cryptic (meaning that it cannot be seen by conventional karyotype and FISH) and so molecular studies needs to be performed to identify this translocation. Also, ALL adults with t(9;22), *IKZF1* deletions, and *JAK* 1/2/3 mutations all have a poor prognosis.

Myeloproliferative Neoplasm

Myeloproliferative neoplasm (MPN) is another hematopathological entity that has many molecular alterations. CML is an MPN that has become the poster child of the impact that molecular diagnostics can have in oncology. CML is defined by the t(9;22) that results from a translocation between the *ABL1* on chromosome 9 and the *BCR* on chromosome 22 forming the *BCR-ABL1* fusion gene on the derivative chromosome 22 that leads to unregulated cellular proliferation. Imatinib, a tyrosine kinase inhibitor, binds to the active binding site of the abnormal tyrosine kinase and inhibits the uncontrolled proliferation of tumor cells with the *BCR-ABL1* fusion genes. Since the discovery of this fusion gene and Imatinib, the survival rate of patient with CML has dramatically improved.

Due to the different breakpoints in *BCR* and *ABL1*, multiple fusion transcripts can result. Despite having multiple breakpoints of the *ABL1* gene at 9q34, due to splicing mechanisms, at the RNA level *BCR* is usually fused to exon 2 of *ABL1*. On the other hand, most commonly, the breakpoints in *BCR* are located in the major breakpoint cluster region (M-bcr) (translated into p210 protein, which occurs most commonly in CML) and the minor breakpoint region (m-bcr) resulting in the p190 protein seen often in ALL patients. More uncommonly, the breakpoint on BCR can be downstream of exon 19, in the micro-bcr (μ-bcr) region, resulting in p230 protein that is associated with the neutrophilic variant of CML.

For patients suspected to have t(9;22), a fusion FISH probe is usually first used to detect the *BCR-ABL1* fusion. This is then followed by RT-PCR to detect the exact fusion that has occurred (p190, p210, and/or 230) due to the prognostics and therapeutic implication of each of these fusion genes. Lastly, quantitative PCR (qPCR) is used to monitor for minimal residual disease in these patients following treatment. In CML, several categories of remission are present. The first is morphologic remission (not seen on morphology), followed by cytogenetic remission (not seen by conventional karyotyping), and lastly there is molecular remission. Based on the National comprehensive Cancer Network (NCCN) guidelines, the amount of *BCR-ABL1* transcript should be measured at diagnosis and then every 3 months. For major molecular remission, there should be a 3 log reduction or 0.1%; and for complete molecular remission there should be a 4.5 log reduction or 0.01%. Due to the need to follow *BCR-ABL1*'s molecular response to treatment across different laboratories, a *BCR-ABL1* international scale was created. For example, in CML samples harboring the p210 transcript, results are expressed as the *BCR-ABL1/ABL* ratio. The *ABL* in the denominator is a housekeeping gene that is run during the PCR reaction. From this, the international scale (IS) value is calculated by (*BCR-ABL1/ABL1* ratio) × 100 × correction factor. To determine the correction factor, 40 reference samples are tested in a clinical lab, the mean differences (Md) are calculated, and the laboratory-specific conversion factor is the antilog (Md). This conversion factor is then specific to the specific laboratory method used. With the IS value, when patients go to different hospitals with different laboratories, the clinician is able to compare the value of the patient's *BCR-ABL1* transcript across the laboratories. Recently, the WHO developed a series of four reference standards to be used as universal gold standards for *BCR-ABL1* transcript quantification. Now, commercially licensed companies manufacture replicates of these standards who then sell them to laboratories. This is further effort to bring standardization to these qPCR values.

Another important gene in MPN is *JAK2* in the Jannus Kinase family which consists of *JAK1*, *JAK2*, *JAK3*, and *TYK2*. Specifically, in 95% of polycythemia vera patients, a *JAK2* V617F mutation is present with the remaining patients having a mutation in *JAK2* exon 12. The *JAK2* V617F mutation is also seen in 50%–55% of essential thrombocythemia and primary myelofibrosis patients. In these two diseases, 20%–30% of these patients also harbor a *CALR* exon 9 mutations and 5%–10% harbor a *MPL* exon 10 mutation. Some less important mutations seen in MPN include *TET2*, *ASX1*, *CBL*, *IDH*, and *IKZF1* mutations.

Lymphoma

Lymphomas also have many molecular alterations that can aid in diagnosis and can serve as prognostic markers. Many of the common mutations are listed in Table 13.2. As in other areas of molecular oncology, common techniques used to identify these mutations include cytogenetic karyotyping, FISH, and PCR.

MOLECULAR DIAGNOSIS AND INHERITED GENETIC DISEASES

Many inherited diseases have been identified such as cystic fibrosis, hemochromatosis, and sickle cell disease. The most common genetic abnormalities that cause inherited genetic disease include point mutation, insertions/deletions, and tri-nucleotide repeats. Usually inherited genetic diseases are tested and identified in a blood sample from a patient; and since these are germline mutations, the problems of tumor heterogeneity are usually not encountered.

A variety of methodologies can be used to identify inherited genetic disease, ranging from cytogenetic, PCR, and sequencing technologies. To decide what methodology to use, the mutation the test is trying to identify must first be identified. For example, in alpha thalassemia, the most common (95%) mutations are due to deletions, so a multiplexed PCR designed to detect these deletions should be used. For the remaining 5% of alpha thalassemia, point mutations are present and for these sequencing type technologies should be used instead. Testing of genetic disease can

Table 13.2 Common molecular alterations in lymphomas

Name	Mutations	Comment
Follicular lymphoma	t(14;18)	Over expression of *BCL2*
Mantle cell lymphoma	t(11;14)	Up-regulates Cyclin D1
Marginal zone lymphoma	t(11;18)	Common in stomach and lung
	t(14;18)	Common in parotid, ocular adenxa, and liver
	t(1;14)	
	t(1;2)	
	t(3;14)	Common in thyroid, ocular adenxa, and skin
CLL/SLL	17p and 11q deletions	Poor prognosis
	13q deletion	Good prognosis
	Somatic hypermutation (ZAP 70 IHC is a surrogate marker)	Good prognosis
Burkitt lymphoma	t(8;14)	Most common
	t(2;8), t(8;22)	Rare variants
Anaplastic large cell lymphoma	t(2;5)	Most common
	t(1;2), t(2;3), inv(2), t(2;17), t(X;2)	Other less common abnormalities
Myeloma	del 13q, 1q+, del 17p, t(4;14), t(14;16)	Poor prognosis
	Trisomy	Good prognosis

be used as a complementary or a primary test to identify these inherited genetic conditions. For example, in patients suspected of Factor V Leiden, coagulation screening tests can be performed initially and DNA analysis can be used as a confirmatory test. However, DNA analysis can be used as the first line test also.

One important issue to consider in inherited genetic testing is genetic counseling. This is important because it is important for the patient to be informed of the implications of the inherited genetic tests before and after they are performed. It is also important to obtain consent before testing because some patients might not actually want to know if they have an inherited genetic disease.

Examples of Inherited Diseases

While there are many inherited genetic diseases, only two categories of inherited genetic diseases will be briefly presented here since most pathologist are not involved in inherited genetic disease testing. A category of inherited genetic disease is triplicate diseases such as Huntington disease, Friedreich's ataxia, Myotonic dystrophy, and Fragile X syndrome. In Fragile X syndrome, CGG repeats are present in the 5′ untranslated regions and they are the most common inherited form of mental retardation associated with tremor/ataxia syndrome and premature ovarian insufficiency. In a wild type patient, 5–44 repeats are present; in an "intermediate" patient, 45–54 repeats are present; in a "pre mutation" patient 55–200 repeats are present; and in a "full" mutations patient, greater than 200–230 repeats are present. Fragile X testing consists of PCR and Southern blot (used concurrently or as a reflex testing) and is performed on patient with mental retardation, and for carrier status.

Another example of an inherited genetic disease is mitochondrial disease. These diseases are inherited from the maternal side. Mitochondrial DNA are circular DNA of approximately 16,000 base pairs and are mainly coding DNA that code for proteins used in the respiratory chain and RNAs. Here, the severity is based on how many normal versus mutant mitochondria are present. Examples of mitochondrial diseases include neuropathy, ataxia, retinitis pigmentosa, Kearns–Sayre syndrome, and Leber hereditary optic neuropathy.

APPLICATION OF MOLECULAR TESTINGS IN DIAGNOSIS OF INFECTIOUS DISEASE

Molecular infectious disease is the most mature field in molecular diagnostics evidenced by the fact that they have the most number of FDA approved tests available. One reason for this is the sheer volume of infectious disease tests being run clinically. For the most part, molecular diagnostics in infectious disease mostly focuses on the identification and monitoring a pathogenic organism (bacteria, virus, or parasite). Many different techniques are used to identify the organisms, including PCR/variants of PCR, sequencing, or probe hybridization. The main advantages of using molecular techniques when compared to traditional microbiology techniques are faster turnaround time and higher sensitivity. Despite these advantages, one major disadvantage is that since the molecular technique is only identifying the pathogen's DNA, it is not able to distinguish whether the pathogen is an active replicating organism or a dead organism. So despite the higher sensitivity for the identification of the organism, the specificity for an active replicating organism is lowered.

Molecular Bacteriology

In bacteriology, traditional culture and biochemical techniques is actually quite good at identification of most bacteria. The main utility of molecular techniques in bacteriology is for organisms that are hard to culture. For example, *Chlamydia trachomatis* is an obligate intracellular bacterium and is hard to culture due to the intracellular nature of this bacterium. Another bacterium is *Neisseria gonorrhoeae* which is a fastidious Gram-negative coccus that is also hard to culture because it is hard to transport. These two bacteria are usually tested together when using molecular techniques. However, as mentioned above, while it has increased sensitivity, it is not specific to active replicating organisms. Due to this, culture is still the gold standard for medical-legal cases that involve *Chlamydia* and *Gonorrhea* infections. In fungus identification, molecular diagnostic's application is similar to certain bacterial identification in that fungal cultures are slow and often difficult to grow.

Another utility of molecular diagnostics in bacteriology is identification of organisms that cannot be distinguished by traditional biochemical test. Different sequencing techniques such as pyrosequencing can be used to aid in the identification of these organisms. The results are compared with the biochemical testing to decide what the organism is and whether to do further testing. To perform pyrosequencing in this situation, a pure bacterial sample is obtained from a culture and after the DNA is extracted, the DNA is run on a PCR (three common regions for all bacteria is amplified), the data is analyzed by comparing with bacteria databases, and the results are sent to the medical director who compares it with biochemical results and gives an interpretation.

Molecular Virology

In addition to identification of viruses with molecular techniques, it is often used to monitor disease and identify resistance to therapy. A good example is in human immunodeficiency virus (HIV), which a virus with a RNA genome that infects and kills CD4+ T-cells. A general algorithm for HIV is diagnosis with antibody and antigen testing followed by measuring the viral load with a quantitative reverse transcriptase-PCR (qRT-PCR) on the plasma. The purpose of measuring the viral load is to help monitor the disease progression and response to antiviral therapy in the patient. Viral load in HIV has traditionally been measured in copies/mL; however, as in *BCR/ABL1* testing, WHO testing standards are now available and so it can also be reported in international unit (IU) when using the WHO standards. One important point to keep in mind is that viral load can increase with other infections and vaccinations, so retesting after 1 month of resolution is warranted in these two situations. Furthermore, antiviral resistance can develop in HIV and so by examining the genotype of the virus with sequencing techniques, a laboratory can look for mutations in the virus that is associated with resistance.

Point of Care Testing

In identification of infectious disease pathogens, turnaround time is a vital issue. The idea is that the faster the results are delivered, the faster the treatment can be altered to most effectively treat the infection. These can result in decrease antibiotic usage, length of hospital stay, development of resistance to therapy, and morbidity and mortality in patients. This is the reason that many different platforms and assays for point of care testing of pathogenic organisms were developed and are also approved by the FDA. Initially, the point of care testing was for single organisms. For example, for *Clostridium difficle*, an assay is available to be used on the GeneXpert platform. This assay can be run 24/7 in the laboratory and can detect toxin B gene.

Due to the advancement of molecular diagnostics in the infectious disease field, many FDA and non-FDA approved point of care multiplexed panels for different pathogens are now available. Many of these panels are syndromic panels that tests for all/most of the bacterial, viral, and/or parasitic pathogens causing a syndrome such as gastroenteritis. Two popular syndromic panels used for gastroenteritis is the Biofire FilmArray GI panel and the Nanosphere Verigene Enteric Pathogen Test. Both of these panels are cartridge-based panels where the user would mix the stool sample with a solution and inject the sample mixed with the solution into the cartridge. The cartridge is then placed onto the platform's machine and through a multiplexed PCR and their own propriety detection method the pathogens are detected. The machine will issue a report on the specific pathogen detected. These types of assays only require 2–5 minutes of hands on time and the machine will run the test in 1–2 hours. The Biofire GI panel tests for 22 different pathogens (bacterial, viral, and parasites) that are common causes of gastroenteritis and the Biofire also has a respiratory pathogen, meningitis pathogen, and blood culture pathogen panel. The Verigene Enteric Pathogen Test identifies nine different common bacteria and viruses that cause gastroenteritis. A Gram-positive blood culture, Gram-negative blood culture, and a respiratory pathogen flex test panel are also available with the Verigene platform. Besides the faster turnaround time and increased sensitivity of these multiplex cartridge-based panels, ease of use and ability to detect multiple pathogens with a relatively low amount of starting sample material are other advantages. However, as with other nucleic acid-based test, the user is not able to distinguish actively replicating pathogens from dead nonreplicating pathogens with these panels.

Future Applications

In the horizon, there are different molecular techniques that offer the same advantage of faster detection of pathogens but can also distinguish between live and dead pathogens. For example, a company called Geneweave has developed particles they named "Smarticles" that are pathogen specific bioparticles designed to bind to a target bacteria. Within these bioparticles, there is a DNA molecule that will be delivered to the specific bacteria and will produce light if the bacteria are alive. So this technology can be used for both detection and susceptibility testing in the future. For detection, a clinical sample can be mixed with a Smarticle for a specific bacteria, and if live target bacteria is present in the sample, a light signal will be produced. If no live bacteria are present in the clinical sample, then no light will be produced. For susceptibility testing, a Smarticles, clinical sample, and an antibiotic will be placed in an assay. If light signal is produced, then the target bacteria are present and not killed by the antibiotic meaning that the bacteria are not susceptible to the antibiotic in the assay. If no light signal is produced, then the target bacteria are not present or are susceptible to the antibiotic in the assay. This type of technology and others will continue to be developed and continue to change the field of molecular diagnostics.

PHARMACOGENETICS

Molecular applications in pharmacogenetics are still in its infancy when compared to its application in inherited disease and infectious disease. Pharmacogenetics is a field that examines response to a drug based on genetic variations in patients. The idea is to use the right drug and right dose for a patient based on their genetic variation in order to minimize adverse drug reactions and to maximize efficacy of the drugs. As mentioned in the molecular oncology section, companion diagnostics is a type of pharmacogenetics that is already being readily used for certain malignancies; however, in this discussion of pharmacogenetics, the focus will be on nononcological drugs.

While an exciting field, several challenges exist in the pharmacogenetics field. One major challenge is that often times many factors besides the patient's genotype are involved. These factors can include age, sex, disease state, exercise status, coadministered drugs, diet, smoking, and alcohol drinking. Another major challenge in pharmacogenetics testing for nononcological drugs is that the genetic tests for these drugs, even if available, are not usually ordered upfront, but is instead ordered after the drug has already been given and after an adverse drug reaction has already occurred. Even if the test was ordered before the drug was given, the results are usually not back in time before the first dose is given. Also, while the FDA has some black box warnings that recommend genetic testing for certain nononcologic drugs, they are not usually required. Hence, dosing of these drugs is usually not based on information regarding the patient's genotype and is dosed solely on data in preclinical trials (dosed based on weight, age, sex, etc.). As this field continues to evolve, the ideal situation is to change clinical practice to test the genotype of patients for drugs before the first dose is given in order to minimize the adverse drug events and to increase the efficacy of the drugs.

Pharmacokinetics/Pharmacodynamics

An important concept to understand in pharmacogenetics is the difference between pharmacokinetics and pharmacodynamics. Pharmacokinetics describes the efficacy of a drug that is influenced by what the body does to the drug through absorption, distribution, metabolism, and excretion of the drug. Pharmacokinetics is traditionally evaluated by measuring the drug and/or their metabolite in the blood stream at different intervals of time. In pharmacokinetics, phase 1 describes the process by which a drug is activated to become functional in the body and phase 2 describes how the drug is inactivated/made more water soluble for excretion. The main group of enzyme that is involved in the phase 1 process is the *CYP450* enzymes. It is important to note that these are not the only genes involved. This group of enzymes includes 57 genes and 59 pseudogenes and is involved in the metabolism and detoxification of many drugs. Some of the commonly involved genes include *CYP3A4/5*, *CYP2D6*, *CYP2C9*, *CYP2C19*, *CYP1A2*, and *CYP2E*. The genes for these enzymes are very polymorphic leading to the variety of responses that the patient can have to the drug based on their genotype for these genes. Due to polymorphism of these genes and/or others, a patient can be categorized as an ultrarapid metabolizer, extensive metabolizer (normal or wild type), intermediate metabolizer, and poor metabolizer. The basic concept is that if the patient is an extensive metabolizer, then a higher dose should be given to the patient; if the patient is a poor metabolizer, then a lower dose or alternative medicine should be given; and if they are an extensive metabolizer, then a normal dose should be given. The process of phase 2 consists of inactivating or making the drug more soluble for excretion and this is usually performed by transferases with enzyme mediated conjugations with substances such as acetate, sulfate, and methionine. The other major concept is pharmacodynamics which deals with the safety of a drug by examining what the drug does to the body. It measures whether the drug is therapeutic or toxic and is measured by biomarkers such as INR.

The nomenclature in pharmacogenetics is unique and will be touched upon here. After the gene is named, such as *CYP2C9*, it is followed by identification of the specific allele. It is the variation between the alleles that lends to the polymorphism of the genes. The allele is first identified by a ⋆ followed by a number and possible an alphabet. The alleles are numbered in the ordered they were discovered and not based upon other numbering schemes. Hence, the ⋆1 allele is the first discovered or wild type allele and is the fully functional allele. ⋆2, ⋆3, so forth are the subsequently discovered alleles and besides defining a specific variant it also represents a unique haplotype of linked SNPs over a region. The alphabet (i.e., in ⋆2A, ⋆2B, etc.) are additional variations in the designated allele that define the SNP and are considered subfamilies.

Commonly Tested Genes

While many genes are involved in pharmacogenetics of various drugs, some common genes tested for are *CYP2C19*, *CYPD6*, *CYP2C9*, *UGT1A1*, and *VKORC1*. Common assays to detect these variations include Roche Amplichip, Luminex Xtag, Autogenomics INFINITI, Hologic Invader, ABI TaqMan, and various laboratory developed assays. Most of these assays are based on PCR technology with their own proprietary detection techniques.

Examples in Pharmacogenetics

Pharmacogenetics can be applied to a wide range of drugs such as Tamoxifen, Statins, Azathioprine, and Tacrolimus. In this section, the application of pharmacogenetics in Warfarin and Clopidogrel, two coagulation drugs, will specifically be presented.

Warfarin is a vitamin K antagonist that is used in anticoagulation. This drug has a narrow therapeutic window meaning that the drug concentration needs to be in a narrow range in body for the drug to work effectively for the patient. If there are too much of the drug, the patient can have bleeding; and if there is too little of the drug, the patient can have thrombosis. Traditionally, Warfarin is monitored by international normalized ratio (INR) and screening of certain coagulation factors (1, 2, 7, and 9) and fibrinogen. It is metabolized *CYP2C9* and targets *VKORC1*, and the genetic variability of these two genes causes 35% dose variability. The alleles that are specifically examined include the *CYP2C9* *2 and *3 which represents a patient that is a poor metabolizer and the *VCORC1*-1639G > A mutation. Other factors in Warfarin dosing include age, height, ethnicity, etc. This is an example of a drug where the genotyping is recommended but not mandated by the FDA.

Clopidogrel is an antiplatelet (anticoagulation) drug where pharmacogenomics testing may be useful. A standard dose is 300 mg followed by 75 mg daily and VerifyNow is traditionally used to measure its effectiveness. However, patients can be poor metabolizers of this prodrug due to allelic variations in *CYP2C19*. Patients with *1/*1 are extensive metabolizers, *1/*2 are intermediate metabolizers, *17/*17 are ultrarapid metabolizers, and *2/*2 are poor metabolizers. For this drug, the FDA has a black box warning that states that there is diminished effectiveness in poor metabolizers, mainly dependent on *CYP2C19* and that this genotype can be used to help in determining the dosing strategy. Patients who are poor metabolizers cannot metabolize the drug to its active form and hence needs more of the drug to achieve the same anticoagulation effect as the standard dosage. An alternative for these poor metabolizers is to use an alternative such as Prasugrel.

KEY POINTS

- Companion diagnostics means pairing a diagnostics test with a drug to predict how well the drug will work on the patient before giving the drug to the patient.
- Knudson's hypothesis refers to a "two hit" theory (based on tumor suppressor genes), where the patient has a mutation (this is the "first hit" and is usually a point mutation) inherited in his/her genome, and when a second mutation (usually a deletion, point mutation, or methylation) is acquired ("loss of heterozygosity"), the patient rapidly develops a malignancy.

- Common tumor suppressor genes include *TP53* located on 17q13, *RB* located on 13q14.1, *APC* located on 5q23, and *VHL* located on 3p26-25.
- Tumor-genesis also occurs by the activation (with a point mutation, amplification, or translocation) of proto-oncogenes which encode proteins that affect cell growth, differentiation, apoptosis, and signal transduction.
- Common oncogenes include *K-RAS* on 12p12, *MYC* on 8q24, and *BRAF* on 7q34.
- A germline mutation is a mutation that is present in an egg or a sperm, is heritable, affects all cells in an offspring, and can cause a cancer family syndrome; and a somatic mutation is a nonheritable mutation that occurs in nongermline tissues such as breast, colon, and thyroid.
- MSI does not have specific therapeutic targets, but it has certain prognostics implication and it is also used to help detect Lynch syndrome (also known as HNPCC).
- MSI occurs when there is expansion or contraction of microsatellite repeat and can be seen when germline mutations in mismatch repair genes *MLH1*, *MSH2*, *MSH6*, and/or *PMS2* are present and also in sporadic colorectal cancer through acquired hypermethylation of *MLH1*.
- Testing *KRAS2* mutations in codon 12, 13, 61, and sometimes 146 in colorectal cancer can be used to determine the effectiveness of *EGFR* inhibitor therapy (e.g., Cetuximab).
- FAP patients have an *APC* knockout mutation and clinically they have multiple adenomatous polyps that have a 100% risk of transforming into colorectal cancer if the colon is not removed.
- Commercial gene panels such as OncotypeDx and Mammaprint are available to predict prognosis and/or help guide treatment decisions in breast cancer patients.
- *HER2* amplification is used to identify breast cancer patients who usually respond well to Herceptin (a monoclonal antibody).
- *BRCA1* (17q) and *BRCA2* (13q) mutations can be performed to predict increased lifetime risk for breast cancer and ovarian cancer.
- *BRAF* V600E is a common mutation in papillary carcinoma of the thyroid, is seen in 70% of tall cell papillary carcinoma, and is also seen in other cancers such as colon cancer and melanoma.
- *RET/PTC* translocation and *RAS* mutations are also seen in papillary thyroid carcinoma, while in follicular carcinomas, *RAS* mutations are also seen along with *PPARγ/PAX8* translocations.

- Codeletion of 1p/19q in grade 2/3 oligodendroglioma shows improved response to chemo-radiation and increase survival.
- *EML4-ALK* translocation, like *EGFR* mutations, occurs more commonly in nonsmokers; however, *KRAS* mutations occur more commonly in smokers.
- In sarcoma, many of the translocations are an objective method to diagnosis the sarcoma; however, these translocations may also have prognostic and therapeutic implications.
- Many diagnosis and classification schemes in the WHO for hematopathology are based on the presence of specific molecular alterations.
- Bone marrow and leukemic blood cells should be collected in EDTA (purple top tubes) for molecular testing and samples for cytogenetic testing should be collected in sodium heparin (green top tubes).
- T-cell or B-cell clonality testing is mainly performed to aid in the distinction of a benign versus malignant neoplasm in an atypical lymphoid proliferation.
- Leukemic cells with translocation of *RUNX1-RUNX1T1* t(8;21)(q22;q22); inversion of *CBEB-MYH11* inv(16)(p13.1q22) or t(16;16)(p13.1;q22); and *PML-RARA* t(15;17)(q22;q12) are AML regardless of blast count.
- Other significant mutations in AML include *FLT3*-ITD which is associated with reduced survival; *FLT3*-TKD with unclear clinical significance; *NPM1* mutation which has a good prognosis if *FLT3*-ITD negative; and homozygous or bi-allelic *CEBPA* mutation that has a good prognosis.
- In ALL patients, many cytogenetic abnormalities have prognostic implications.
- CML is defined by the t(9;22) and Imatanib, a tyrosine kinase inhibitor, works extremely well on patients with this translocation.
- *JAK2* V617F is seen in 95% of polycythemia vera patients and in 50%–55% of essential thrombocythemia and primary myelofibrosis patients.
- Lymphomas have many molecular alterations that can aid in diagnosis and can serve as prognostic markers.
- The most common genetic abnormalities that cause inherited genetic disease include point mutation, insertions/deletions, and tri-nucleotide repeats.
- One important issue to consider in inherited genetic testing is genetic counseling because it is important for the patient to be informed of the implications of the inherited genetic tests before and after they are performed.

- In Fragile X syndrome, CGG repeats are present in the 5′ untranslated regions and they are the most commonly inherited form of mental retardation associated with tremor/ataxia syndrome and premature ovarian insufficiency.
- Mitochondrial diseases are inherited maternally and examples include neuropathy, ataxia, retinitis pigmentosa, Kearns–Sayre syndrome, and Leber hereditary optic neuropathy.
- The main advantages of using molecular techniques when compared to traditional microbiology techniques is faster turnaround time and higher sensitivity; however, the assays are not able to distinguish whether the pathogen is an active replicating organism or a dead organism.
- Many FDA and non-FDA approved point of care multiplexed panels for different microbiological pathogens are now available.
- Pharmacogenetics is a field that examines the response to a drug based on genetic variations in patients so that the right drugs and the right dose of drugs can be given to a patient based on their genetic variation in order to minimize adverse drug reactions and to maximize efficacy of the drugs.
- A challenge in pharmacogenetics is that besides the genotypes of the patients, environmental factors can also contribute to the efficacy of a drug for a particular patient.
- Commonly tested genes in pharmacogenetics are *CYP2C19*, *CYPD6*, *CYP2C9*, *UGT1A1*, and *VKORC1*.

REFERENCES

[1] Harris TJ, McCormick F. The molecular pathology of cancer. Nat Rev Clin Oncol 2010;7:251–65.
[2] Baker SJ, Kinzler KW, Vogelstein B. Knudson's hypothesis and the TP53 revolution. Genes Chromosomes Cancer 2003;38:329.
[3] Bunnik EM, Schermer MH, Janssens AC. Personal genome testing: test characteristics to clarify the discourse on ethical legal and societal issue. BMC Med Ethics 2011;12:11.
[4] Lewandowska M, Hybiak J, Domangala W. Concordance of KRAS mutation status between luminal and peripheral regions of primary colorectal cancer: a laser capture microdissection based study. Pol J Pathol 2016;67:13–18.
[5] Waller A, Findeis S, Lee MJ. Familial adenomatous polyposis. J Pediatr Genet 2016;5:78–83.
[6] Miline RL, Antoniou A. Modifiers of breast and ovarian cancer risks for BRCA1 and BRCA2 nutation carriers. Endocr Relat Cancer 2016;23:T69–84.

INDEX

Note: Page numbers followed by "*f*" and "*t*" refer to figures and tables, respectively.

C

D

E

O

P

Q

R

S

T

Y

Z

Printed in the United States
By Bookmasters